The Experimental Method

The Experimental Method

A text for students of
Engineering and Science

Edited by R. K. Penny
*Professor of Engineering Design and
Production, University of Liverpool*

Contributors to the text are:
J. C. Gibbings, S. L. Dixon, D. G. Moffat,
A. K. Lewkowicz, F. Drabble, R. Shaw,
R. K. Penny

Longman

LONGMAN GROUP LIMITED
London

*Associated companies, branches and representatives
throughout the world*

© Longman Group Limited, 1974

First published 1974
ISBN 0582 44351 2 cased
ISBN 0582 44352 0 paper

Library of Congress Catalog Card Number: 73-86103

 PRINTED BY Unwin Brothers Limited
THE GRESHAM PRESS OLD WOKING SURREY ENGLAND

Produced by 'Uneoprint'
A member of the Staples Printing Group

Contents

	Page
Acknowledgements	vi
Preface	vii
1 **Introduction** *by R.K.Penny*	1
2 **Experimental Method and Procedure** *by R.K.Penny*	6
3 **Dimensional Analysis Applied to Experimentation** *by J.C.Gibbings*	21
4 **Instrumentation** *by S.L.Dixon and D.G.Moffat*	33
5 **Errors in Experimentation** *by A.K.Lewkowicz*	73
6 **Analysis and Interpretation of Results** *by F.Drabble*	80
7 **Communication** *by F.Drabble*	116
Appendix 1 *by J.C.Gibbings*	128
Appendix 2 Outline of error analysis *by A.K.Lewkowicz*	130
Appendix 3 Notation, units, laws, formula, properties, and constants by *R.Shaw*	153
Index	177

Acknowledgements

We are grateful to the following for permission to reproduce copyright material:

The Biometrika Trustees for their table of the Gaussian integral; G. P. Putnam's Sons and Hodder and Stoughton Ltd for an extract from *Gypsy Moth Circles the World* by Sir Francis Chichester, 1967.

Preface

In 1969 a committee was formed within the Department of Mechanical
Engineering of Liverpool University to discuss the objectives of experimental
work in its undergraduate curriculum. The committee was aware of fashionable
trends towards 'open-ended' experiments and the limitations of some experiments,
commonly in use, that had remained unchanged for years. After many long discussions the committee concluded that there was room for all types of experiment
ranging from the short demonstration to the longer project which would be undertaken in the final year of the course. As a result a gradation of open-endedness
was decided upon wherein it was considered necessary that undergraduates should
be exposed to a process of:

        (a)   acquisition of measurement skill,
        (b)   problem solving in the laboratory,
        (c)   undertaking a major project.

In the three-year course, the first year involves item (a), the second items (a)
and (b), while the third year is concerned with (c). In this scheme it was envisaged
that the development of students' responsibility and initiative as well as the use of
judgement and communication skills could best be achieved.

The committee further concluded that it was necessary to provide a
number of background lectures very early in the course on such topics as method,
planning, instrumentation, errors analysis and interpretation, and communication.
At the same time suitable texts to cover these aspects were sought. Although
many excellent books were found to be available none seemed to be completely
suitable. Those available were either too advanced or too specialised or some
aspects just were not covered at all. To rectify this situation each committee
member undertook to write brief notes on each of the topics mentioned with a view
to using them as student handouts that complemented lectures on the topics. Since
that time these handouts have been refined and expanded in the light of subsequent
experiences in lecture classes. Finally, it was decided to coordinate the individual
efforts to form this short book.

The book is not intended to be completely comprehensive nor is it a catalogue of experiments. Its accent is on methods of approach to the various phases of experimentation. As such it should be useful to students of the natural sciences as well as to those of engineering. It is an elementary text intended to bridge the gap that often exists between schools and university courses; it may therefore be appropriate for use by school pupils.

1 *Introduction*

1.1 Object of the book

The object of this book is to provide students with a viewpoint on experimentation. It is only a viewpoint, since as the experimenter becomes more skilled he develops his own methods and healthy scepticism of the ways of others. He also learns how to process seemingly unimportant scraps of information gleaned from his experiments, often following a 'hunch' or as a result of a theoretical possibility. When he is first starting, though, it is better for him to have *some* guidance on method and planning, basic instrumentation, errors in results and their interpretation and how to communicate his findings to others. This is what the book is about.

1.2 The need for experimentation

In an age when abstract thought is fast becoming regarded as the highest calling of man it is often forgotten that most thought is based on observation and that observation invariably involves doing, or being involved in doing. 'Doing' here means experimentation and experimentation is vital for progress in any discipline where information is lacking. A visit to any engineering works quickly reveals that experimentation is going on in a wide variety of activities. Prototype testing is done to verify that performance requirements are being achieved or that a part is strong enough. Parts in service sometimes fail and it is often necessary to conduct further component tests, perhaps during maintenance periods, to find out why service conditions encountered were different from those that were anticipated. Perhaps a new alloy has been developed and it is necessary to verify that its properties are in fact better than previous ones. Clearly the whole process is a very expensive one and if the information could be obtained through the application of thought and the solving of equations then this, the cheaper route, would be adopted. Of course, an integrated procedure usually goes on which involves application of physical principles, solving of equations which state these principles mathematically, design based on analysis, followed by production of the part or process in question. At the end of the day though the achievement of the most successful and reliable outcome will only be realised if all the imponderables and unforeseen facets of the problem area are physically tested.

Precisely the same story holds of any true research activity; reinforcement of ideas and postulates comes after considerable experimentation, observation, testing of the ideas, re-postulation and re-testing.

1.3   Implementation

Experimentation can be extremely laborious and for the best results of widest applicability, derived with economy, methodical approach and careful planning are vital. These are the subjects of Chapters 2 and 3. Chapter 2 may seem to be no more than a statement of the obvious—who would consider doing anything unmethodically? It is surprising to find in educational circles and in industrial laboratories, however, that obvious questions often have not been asked. What is the real aim of the experiment? What is already known about the problem area? Which steps should come first? Why? Which instruments are available or are needed? What are their limitations? What other resources are needed? What do the results mean? And so on. A naphazard approach clearly is not going to be economic or fruitful, and the novice soon discovers this to his peril if he does not adopt some sort of method. Chapter 2 makes some suggestions that can be tried by the student and the suggestions can be augmented as he goes along and as he gathers experience. Chapter 3 outlines some aspects important in planning which can be aided by dimensional analysis. There are many sophisticated tools available for planning of experiments but generally they are beyond the novice. Dimensional analysis is not and it started decades ago primarily as an aid to experimental strategy. It has achieved greatest popularity in the fields of fluid mechanics and heat transfer but over the years has been considerably expanded to other fields. It is unfortunate that, instead of the principles of dimensional analysis being used to enable experimenters to improve their techniques, the subject is often presented as a historical curiosity. In fact it is a powerful and rapid way of deriving some basic functional relationships without the need for complex theoretical deliberations. It is hoped that Chapter 3 goes some way to countering the pedagogical approach and opens up a means of *utilising* this important branch of knowledge to full advantage.

All forms of experimentation involve instruments. The range of instruments can vary from, say, a simple rule to an electron microscope and no book could describe all possibilities. Chapter 4 devotes some space towards the simpler end of the scale just mentioned and at the same time provides an introduction to elementary concepts of sensing, transduction and processing of such quantities as length, flow rate, temperature and others. It is thought that in this way the greatest educational value results from teaching about principles of *some* instruments and instrumentation rather than attempting to describe a multiplicity of instruments. This stems from the belief that instruments should be treated as part of the *measurement* problem. The student, if properly educated, will come to realise the proper position of an instrument in his problems. As he grows in his appreciation he will be able to design his own instruments if he does not easily have to hand ready made devices. Through the use of elementary instruments early in his course he will appreciate such things as accuracy, response, interference and other basic characteristics. Later during his education there will be ample opportunity to utilise high-speed, modern and involved instrumentation once a scepticism and proper respect for the early ones have been developed.

No matter what the experiment, some form of data will always be a product. In the simplest experiment these data almost certainly will contain errors and certainly should be deserving of interpretation. These topics are the subjects of Chapters 5 and 6.

A study of potential errors and uncertainty is part of the planning of methodical experimentation. No matter how carefully this is done errors will arise, however small. If these are not critically examined much loss of time and money will result; even worse, technical untruths resulting perhaps in dangerous failures could follow. Chapter 5 examines the different types of error that the careful experimenter will want to try to quantify and assess the importance of; these might include human errors or instrument errors which can arise in random or systematic fashions. The nature of precision, error and uncertainty is reviewed in some detail and in order to make the book reasonably self-contained the basic mathematics of statistics needed at the elementary level has been included in Appendix 2. To the beginner the mathematical content may seem rather heavy. From this there can be little escape, however, and with illustrations by examples in Chapter 5 followed by practical use, experience shows that most students are capable of appreciating the need for understanding and generally they have the ability to master the mathematical concepts involved.

Statistical analysis is not the only form of data checking and other necessary types of interpretation are discussed in Chapter 6. In any case, comparatively few engineers feel completely at ease with statistical concepts; statistical inferences appear to lack precision in spite of the fact that the subject is most useful in tests which themselves contain inexactitudes. Engineers are more at home with graphs which can often represent all the data concisely and with no manipulation. Unfortunately graphical methods have many pitfalls if not used properly. Poor curve fitting can lead to fallacious theories unless suitable checks are made; for example, through balance equations: mass, energy, momentum, current and so on. False conclusions regarding locations and magnitudes of maxima and minima can easily result from bad graph plotting based on sloppy tests for significance or rejection. Extrapolation or interpolation is more than just extending the line that appears to fit a few points in a restricted region to regions far removed by the application of a straight edge to those points.

1.4 Reporting

There are many ways in which the work resulting from experimental effort can be reported. Those who have performed the work may be expected to report to different people in different ways. Masses of detail are unlikely to be important to the administrator who must make decisions about assigning more resources, if they are needed. At the same time *sufficient* information has to be conveyed if the right decision is going to be made. On the other hand if the results are to be presented for close scrutiny by theoreticians—who perhaps requested the experiment—or for publication in an international journal, the work will stand or fall by the detail that is presented. The experimenter has therefore to be very skilled not only in getting his information but just as importantly in the manner of presentation. He has to put himself in the position of the person with whom he is trying to communicate. The sparing but adequate use of good English and clarity of thought in data presentation and analysis are vital whether the report is by written means or oral. All of this is an act of 'doing', just like the experiment itself, and much patient practice is required. The aim of Chapter 7 is to give some guide lines for the student to use when he comes to the stage of communicating to others what he has done.

1.5   The use of this book and others

It has been said several times in this Introduction that the book is not designed to meet all eventualities. As his expertise grows the experimenter will want to refer to advanced texts on specialised aspects; some of these advanced texts are listed at the end of this chapter; references and further reading lists can also be found at the end of each chapter. The references given are by no means exhaustive and the diligent student will want to use libraries and other sources perhaps to seek more. As he matures in his approach to experimentation and other aspects of his discipline he will come to realise he will readily accept that a literature search is a vital first step that can save a lot of time and from which many useful ideas can be extracted. He must thus become used to the idea of using his library facilities to the full; using abstracts, keeping up to date with reviews and journals, following up references from these, using design manuals and so on.

Chapters in the book do not necessarily have to be read in sequence although the sequence given is a logical one. For the beginner it is perhaps better that most effort be put initially into reading Chapters 2 and 3 so that the right attitudes towards method and planning are formed. As a result of *doing* some experiments the student with initiative will want to add to the basic table (Table 2.1) given in Chapter 2 which outlines *some* aspects of strategy. It is unlikely that the examples given in Chapter 3 on grouping of variables will fit those at hand. Again, the initiative of the student is needed to try his hand for his own problem; the scheme will not always work.

Chapter 5 is probably deserving of separate study in conjunction with Appendix 2 and perhaps the student will require backup for this from the lecturers in his course on statistics. Certainly he should not be impatient because he cannot easily master the mathematics involved. There are available these days computer programs that will quickly perform statistical analysis of masses of data. The computer aspect has not been discussed in the book for the simple reason that, with any computer program, the background must first be understood; computer output is, at best, only as good as the input.

With other chapters full value will only be gained by practice and constant reference. Thus the book can almost be used as a manual during the course of the experiment and afterwards during the process of interpretation and reporting. Useful data, facts and figures are presented in Appendix 3 which can be used as a reference source at all times.

References

1. D. R. COX. *Planning of experiments*. John Wiley, New York, 1958.
2. J. P. HOLMAN. *Experimental methods for engineers*, McGraw-Hill Co., New York, 1966.
3. P. W. BRIDGMAN. *Dimensional analysis*, Yale University Press, New Haven, Connecticut, 1931.

The Engineering Index Annual, Engineering Index, Inc., New York.

·*R & D Abstracts*, Technology Reports Centre, Department of Trade & Industry, Orpington, Kent.

Current Contents, (Engineering & Technology). Institute for Scientific Information, Pennsylvania.

2 *Experimental Method and Procedure*

2.1 Method

Experiments usually differ in aspect but generally they all follow the same basic form, i.e. they are subjected to a sequential pattern of planning, implementation and evaluation. This we term *Experimental method*. As with analytical work, a method which allows for the formulation and solution of problems is just as important in experimental work. Each of the steps in the sequence requires that the experimenter always questions his *motive* before proceeding to the next step if successful and economic conclusions are ever to be reached. Unlike analytical work, however, in which it is usually possible to proceed to a unique answer through a single path, it is seldom possible to perform a meaningful experiment this way. As with design work, an experimental programme usually consists of a series of experiments each as part of an iterative process which properly combine with theoretical and analytical tools. The flow diagram in Fig. 2.1 shows schematically *one* method which could be helpful in proceeding to a meaningful outcome.

The diagram may appear to be somewhat trite and certainly is not foolproof. Nevertheless it assembles the essential characteristics and arranges them in an articulated form which then enables sensible interaction and repetition once various steps have been taken. Often there will be no 'existing information' and the box with that heading will first be filled when the experimenter has undertaken some preliminary tests to establish the effect of a range of variables or the likely time scale of a fully fledged programme in which existing or new apparatus has been tried. As a result of these preliminary tests the experimenter now has *some* information about his problem which, together with the initial experience of equipment, will help to generate new ideas and even to re-formulate his objectives. This discipline, sensible programming of events and self-questioning by the experimenter, are vital if objectives are to be fulfilled economically (in time or money). Particular skills which have to be learned in order to implement what comes under 'test sequence' are discussed later in this chapter and in other chapters of the book.

Until now no concepts which have restricted the discussion to any

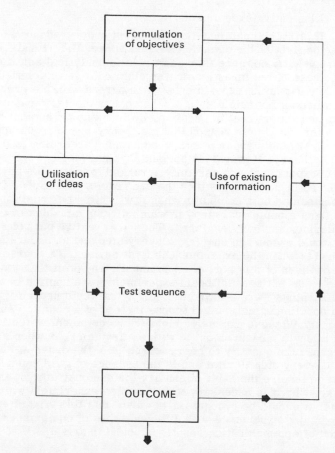

FIG. 2.1   Essential details of experimental method

particular discipline have been introduced. Indeed it is a noteworthy fact that the field of experimentation is one of the more common links between scientists and engineers. This is not surprising since it is just one of the ways of seeking information which is otherwise not available.

So far as engineering students are concerned laboratory work has at least four objectives:

1. For experiments designed to illustrate points covered in lectures.
2. For experimental investigation.
3. For design and synthesis.
4. As training in measurement techniques.

Illustrative experiments

It is, unfortunately, a fact that most undergraduate experiments usually fall into this, the first of the categories listed above. Unfortunate, because such experiments are usually no more than analogues to analytical solutions. One is reminded of the case of the investigator who claimed to have verified a theory for tide motion by performing experiments which agreed with his predictions. The theory was hailed as invincible until another experimenter measured the *actual* motion of the sea water tract in question. He found no agreement with the first man's theory for the simple reason that the theory was a gross approximation to reality. The first man's experiments were devised according to the assumptions included in the theory. However, important physical features were missing. If all features of an experiment are carefully arranged to match the physics and constraints included in the analysis then the two results will agree. Such experiments are better performed on a computer when it will often be possible to gain wider conclusions from a numerical study in which all the variables of the problem can be quickly and inexpensively explored. The only creative part of such an experiment which would encourage enquiry at every step and an understanding of experimental method is when the experiment is first devised. Of course the student can always *enquire* even of this sort of experiment. Why do it that way? What would happen if the loads in the usual beam experiment were applied in a different way or the beam supports were allowed to sink? What would happen if the beam to be tested had a non-linear material response instead of a neat and easily analysed linear material? Why use expensive measuring equipment which is capable of discriminating a length measurement to within a few microns when all that is needed is a rule because the quantity to be measured is of the order of centimeters? Scepticism at every step is vital and is the essence of good experimentation. If all else fails to inspire the habit of enquiry the demonstrator of an experiment might be asked why it is necessary to perform an experiment which has a predictable outcome. It should be recognised, however, that this type of experiment is instructive in gaining experience with different types of equipment while good practice in the art of communicating results is bound to arise.

Experimental investigation

This enables greater appreciation of all aspects of a problem to be gained in the absence of specific knowledge appropriate to it. Following Fig. 2.1 we may have formulated an objective but unfortunately our task may be beyond analysis and if our task is a new one there will be either very scanty or no existing information. As we have already remarked, we can perform our first experiment, often crudely, from which we generate ideas and add to our store of knowledge. A clue from the first experiment may make a first simple analysis possible with the consequence that a second, more refined, experiment can be planned. Alternatively, existing information may be found to be useless, in which case it should be discarded. According to Gordon [1]* the achievements of the Wright brothers were made

---

* Figures in square brackets refer to bibliographical references at the end of chapters.

possible because they learned early in their investigations how much they still did not know about aerodynamics and how much of the information on which they had been basing their work was inaccurate. Apparently they commented that 'having set out with absolute faith in the existing scientific data we were driven to doubt one thing after another until finally, after two years of experiment, we cast it all aside'. In overcoming the problems of flying (which they accomplished in only four years) they discovered or re-discovered many basic relationships.

Experimental investigation is thus often the only real way to gain the information needed about a problem. When it is, it is usually the last resort because it can be extremely expensive and time-consuming and a method needing all of the skills of an experienced experimentalist to extract the most meaningful results is vital. At all times the main objective must never be lost sight of—in the case of the Wright brothers it was to fly and not to investigate aerodynamic phenomena except in so far as they affected flying. It is easy to be sidetracked into interesting avenues of research until these become unrecognisable from the main objective.

Design

So far as design is concerned, experiment is often invaluable in helping to assess the effects of simplifications made in synthesising complex parts and behaviour to a conceptual level which is capable of mathematical representation. For example, the behaviour of an aeroplane wing spar could be treated, in a simplified form, as a cantilever beam by a trivial analysis. Such an analysis would be useless in realistically assessing the effects of cutouts in the spar or of buckling of the skins. Although it is not always recognised, analysis is usually only capable of describing overall behaviour and often not important details. It is upon the details, however, that the success of most designs depends. Experiment plays its part in verifying not only that the details may be practical possibilities but also that their presence does not impede the main function of the part in question. Another way in which experimentation plays its part as an aid to design is in the choice of a material for a part. Provided that a suitable analysis has already been verified as appropriate, it is often possible to correlate the behaviour of a complex part with a very much simpler laboratory test piece. The laboratory test piece can then be used as the vehicle for comparing different materials, the final choice of which will be dictated from a variety of properties such as strength, toughness, corrosion resistance, weight, cost, machinability and so on.

Generally speaking the feasibility of a design will depend on some imponderable features which could make the difference between success and failure. Does the design work at an assumed efficiency? Will it have the required lifetime? Will the parts fit together to form the whole as expected? Is there an optimum production route? How does the intended user react to the proposed design? and so on. In the absence of any better method of assessing these imponderables experiments performed on a mockup or a physical or mathematical model can put the engineer in a better position of progressing to the stage of detailing his design.

Measurement technique

Soundness of experimental practice is as important to an engineer as his grounding in analysis; in fact the two are inseparable. One of the skills needed in experimental work is in the use and choice of instruments. Thus training in measurement techniques is vital in order that a critical awareness of the value of experimentation and the means of assessing paths of experimentation can be properly developed. Again, a sense of enquiry is essential because when faced with choosing an instrument to do a given job the student must ask of the range of variables to be covered, the effect of environment, whether dynamic effects are important, whether ageing has occurred, how to calibrate, and so on. If a standard instrument is not available can another be utilised? Is it possible to devise one easily and cheaply from scratch? At every stage instruments and techniques should be viewed with scepticism, standards should be regarded with suspicion, calibration charts checked and re-checked.

Whatever the type of experiment, whether it be of any of the types classified previously, a methodical approach is possible and highly desirable. We can train ourselves to account for effects of environment which might occur in random fashion, to plan sequences, how to evaluate and assess the importance of errors, to check and cross check and to communicate our results in an easily understood and orderly fashion. At no time must we generate new concepts based on questionable data and certainly we must not miss the obvious even after lengthy and unnecessary experimentation. In Table 2.1 [2] some of the detailed steps in the Experimental Method have been summarised. The sequence of steps given there is not foolproof and it remains for the student to apply his powers of *logic* and *reasoning* to each step and to improve on the Table as he gains experience. Above all there must *always* be a *motive* for the experiment which results in *action*. The action will be to perform analysis in the light of new data. The self-questioning and enquiry needed during this process is of vital importance in all activities not only in experimentation. An engineer must be able to create and to innovate and he can only do this if he has imagination, a few skills and the willingness to subject himself to self-criticism.

TABLE 2.1   Possible Questions and Actions as aids to methodical experimentation.

STEP	POSSIBLE QUESTIONS		POSSIBLE ACTION
1. Object	What is the object of the experiment?		Find out and state unambiguously.
2. Variables	(i)	What are all the variables?	Preliminary tests
	(ii)	Which are most important?	
	(iii)	What are their ranges?	
	(iv)	Are all variables independent?	
	(v)	Can the variables be reduced by grouping?	Dimensional analysis

TABLE 2.1 Continued

STEP	POSSIBLE QUESTIONS	POSSIBLE ACTION
3. Equipment and environment	(i) Is a special environment necessary?	
	(ii) What previous work has been done?	Literature survey
	(iii) What standards exist, if any? Are they relevant?	Consult codes of practice.
	(iv) What equipment is necessary?	Design test rig.
	(v) What is available?	Use existing equipment if possible.
	(vi) Can tests on simplified models be considered?	
	(vii) Which types of test piece could be considered?	Design test piece. Perform sorting tests.
	(viii) What are the important features of the test piece?	
4. Measuring instruments	(i) What ranges should be considered?	Decide on basis of expected variable range.
	(ii) What accuracy is required?	Decide on relative importance of variable.
	(iii) Which instruments are available?	Use existing equipment if possible. If not design new ones.
	(iv) Are those instruments satisfactory?	Calibrate and do preliminary tests.
5. Procedure	(i) What sequence should be used in varying parameters?	
	(ii) Which tests give simultaneous information on several variables at once?	Plan test procedure.
	(iii) May qualitative observations be important?	Make notes or tape recordings of observations during tests. Devise model tests.

TABLE 2.1 Continued

STEP		POSSIBLE QUESTIONS	POSSIBLE ACTION
6. Evaluation of test results	(i)	Are the results reliable?	Perform cross-checks.
	(ii)	What relationships exist between variables and do they have significance?	Plot results in different ways.
	(iii)	Do the dimensions check?	Consider statistical methods of correlation.
7. Presentation of results	(i)	Which are the results of main significance?	Identify and emphasise in report. Consider graphical presentation, non-dimensional plots, empirical formulae, curve fitting.
	(ii)	How can the results be best presented?	
8. Conclusion	(i)	Do the tests satisfy original object?	
	(ii)	If not, where is the discrepancy and is it important?	Identify discrepancy. Evaluate importance. Propose new tests.

2.2   Procedure

There are no hard and fast rules for proceeding with an experimental programme. Experimenters build up their own expertise and a 'feel' for the job in hand and as experience grows the obvious gets done as a matter of course. This experience is often gained the hard way and usually at great cost and there is no harm therefore in pointing out the obvious, particularly for beginners.

Test sequence

This is most important because it is here that most economies can be made. Decisions have to be made regarding the relative importance of the independent variables thought to govern the outcome of experiments and hence the number of experiments and the order in which they should be performed. At the same time sufficient experiments have to be performed so that the effects of variations due to technique, defects in the apparatus and so on can be properly accounted for.

Although there are mathematical techniques available as an aid to the design of experiments these are usually beyond the scope of the beginner. All that can be given here are some basic guides. So far as the *sequence* is concerned, much depends on whether the phenomenon being investigated is *reversible* or *irreversible*. By reversible we mean that if the independent variable is increased or decreased the measured response follows the same path. An everyday example of a reversible test is in the tuning of a radio. A desired station can be achieved by first going through the station, stopping, going back through and so on, gradually getting nearer and nearer by approaching from both sides until the most satis-factory tuning has been achieved. Often reversible processes are linear in their response and thus regular increments or decrements can be applied. The spacing of test points can be elongated as the tests proceed and trends have been establish-ed. As an example, the determination of the elastic modulus of a rod tested in ten-sion could be considered. In such an experiment sufficient guidance from previous work is available to tell us that stress is proportional to strain. The constant of proportionality between the two cannot be calculated but if measurements of strain resulting from changes in an applied stress are made the constant can be deduced. At the beginning of the experiment the strength of the material is usually unknown so that stresses are initially applied in small increments; points 1 and 2 in Fig. 2.2 may represent two such increments. If we were to connect these points and the origin with a straight line 0-a we could easily be misled into thinking we had

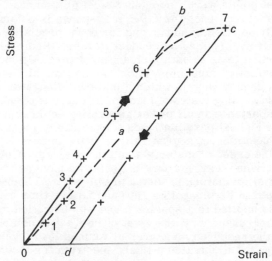

FIG. 2.2   Determination of the elastic modulus

achieved our objective once the slope of the line is calculated. However, the apparatus used may not have 'bedded down' or warmed up so that a few more incre-ments of stress are applied—points 3 and 4. A trend is now established which leads the experimenter to doubt the validity of point 2. More increments, points 5 and 6, are applied to verify the trend. A final increment to point 7 causes the material to yield; the material is no longer in the elastic region except during unloading. The final point is thus rejected and the line of best fit, 0-b, is drawn through the initial points. The slope of this line gives the required elastic modulus. As a final check the knowledgeable experimenter would now reduce the stress in increments to obtain the line c-d which should be parallel to 0-b.

The preceding description is necessarily a simplified account of a tensile test. Many snags await the experimenter even in such a basic and elementary property determination, and even here a *random* testing sequence could be advisable as a means of averaging out apparatus defects, variations due to the environment, experimental observations and batch-to-batch differences. Random sequences are more usually applied to *irreversible*-phenomena, i.e. those processes whose future progress depends on their current and past states. Many examples of this exist in materials testing, the best known of which is probably in fatigue. During the fatigue test the material becomes more and more 'damaged' until it finally breaks; the amount of damage accumulated in the past dictates the load-carrying capacity in the future and the damage cannot be repaired.

Test levels and test point spacing

From the discussion in the last section it was noted that both the spacing and the level of test points were important. Most phenomena confronted by engineers are irreversible but once the *level* of response is obtained the spacing of test points can easily be fixed by arranging for equal precision over the range. The level can be usually ascertained by conducting a small number of preliminary, often rather crude, sorting tests to establish an enveloping region within which to concentrate the main effort of experimentation. As an example let us consider the creep test. Creep is a phenomenon in which a material elongates progressively with time, even when sustaining a steady load, until finally it breaks. It assumes enormous practical importance in all power generating equipment where parts are required to sustain the highest possible loads for the life of the plant without any mismatch of its components or, of course, failure of the material. Theories are not capable of predicting the creep of materials and the only way of obtaining data is from laboratory tests. When testing a new material we have no idea of its capabilities except from similar materials tested previously. We know that the response is highly non-linear both in its elongation rate and failure time but we have no idea how much so. A load is applied to a specimen and its elongation measured. The load is progressively increased until the level of interest is established. New tests are planned and the response measured at various load levels in order to obtain a spectrum. Fig. 2.3 illustrates a likely spectrum for a series of creep tests. Load p_1 was applied first and gave rise to failure in time t_1, which was known to be much shorter than the required design time. A second test at load p_2 was discontinued once the design time t_d had been exceeded. The *level* of test loads could now be taken to be between p_1 and p_2 and more tests performed at loads p_3, p_4, p_5, etc., to fill in the gaps within the shaded area of Fig. 2.3.

In determining the form of the elongation-time curve the spacing of test points is of paramount importance. Because of the highly non-linear initial creep elongation, measurements of elongation after equal time intervals could lead to missing the important first part of the curve shown by the broken part of the line in Fig. 2.4(a). If the time to achieve equal increments of elongation (Fig. 2.4(b)) were measured the complete curve would be obtained. Such a procedure itself is linked with the level of response because we would not know *when* first to measure

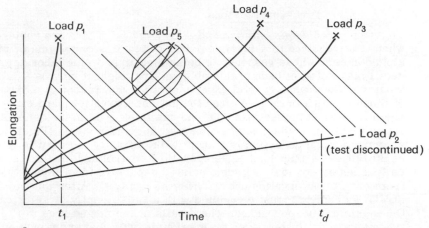

FIG. 2.3   Typical creep test results

the elongation. Thus, having achieved the points of Fig. 2.4(a) from which the
level is measured the spacing in a repeat test could be adjusted to obtain the required data after initial loading. This spacing could be progressively elongated
until it is noted that the elongation *rate* increases as failure of the material (the
cross-shaded region of Fig. 2.3) is approached.

　　　　The description of creep testing which has been given is also greatly
simplified. There has been no discussion of 'scatter'—the experimenter's curse,
the experimental difficulties of measurement, storage and collation of data for
periods up to ten years for a single test and so on. Nontheless, in principle, the job
is approached in the manner described. In the broad description given some important details were omitted and these are worthy of brief discussion.

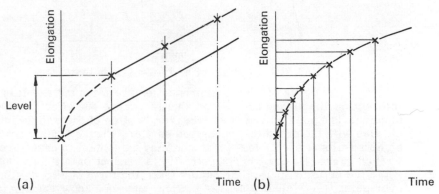

FIG. 2.4(a)(b)   Spacing of points during a creep test.

Procedural actions

Maxima, minima and other details. There are numerous instances when it is required to determine points of maxima or minima or other details of a phenomenon whose response is being measured in the laboratory. The particular detail may not have been predictable; on first sight it may be a quirk of the test. Whatever the reason careful experimentation and reduction of data can reveal any of these points as required given the patience, creativity and experience of a good observer. The simplest example one can think of is in the determination of an upper yield point which can be observed in certain steels (Fig. 2. 5). Although such an experimental deduction is fraught with difficulty in that it is liable to extreme sensitivities to axial loading, temperature control and loading rates, the lessons to be learned are typical. At some point, a departure from linearity will be detected from careful observation and *'plotting-as-you-go'*. After this point much smaller stress increments must be applied with a subsequently larger strain response as the maximum is approached. The details of the ringed area in Fig. 2. 5 can be traced quite accurately even though the absolute maximum may not quite be achieved through the test. Less careful testing and control would not reveal the detail and more frequently the path shown by the broken line in the insert figure would be achieved.

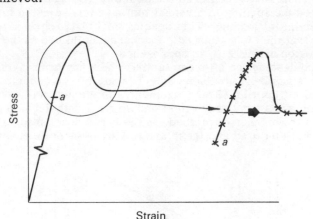

FIG. 2. 5   Details of a stress-strain test

Curves of physical properties are often the result of two or more interacting effects and show slope changes where one effect predominates over another. The point or region of takeover is often of theoretical interest. An example lies with the creep test described earlier. Figure 2.6(a) shows schematically some results of such a test which one may be content to draw through with a French curve. No obvious discontinuity is present in this form but the same data replotted on logarithmic time scales takes the form of Fig. 2.6(b). The change-over point is now clearly defined but how does one know whether to use linear scales, logarithmic scales or whatever? To this question there is no simple answer.

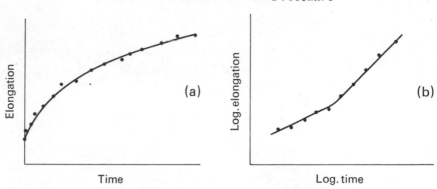

FIG. 2.6   Illustration of range changes

The only recommendation is to try different combinations—*experiment with the observed data*. Some guidance may be available from theoretical considerations but in any event the cunning of the experimenter is all-important.

    Scatter. The subject of scatter of results has been mentioned already. If the experimenter is not careful he can miss the obvious or obscure some feature. He must use his powers of observation to eliminate such features as obvious instrument malfunction and the rejection of stray points but he must keep open at all time his basis for rejection. Statistical methods of reduction of data, cross-plotting techniques, re-plotting to linearise a response, are all techniques which are expanded in greater detail in other chapters. At the same time the level and spacing of test points is very important. A typical plot depicting scatter and the potential dangers of missing important details of a test is shown in Fig. 2. 7(a) where the points from many tests obtained from repeating, over and over again, the stress-strain response of a material. To the uninitiated such differences in such a seemingly, simple test may be exaggerated. Even if the most sophisticated machine is used for the testing and if the most elaborate precautions for temperature and loading rate control are made the feature of a slightly bent or misaligned specimen (often by no more than a few millimeters) could produce such results.

    Replotted in Fig. 2. 7(b) where the tests are distinguished from each other by different symbols, definite trends can be extracted but most important of all is the observation that one test is different from all others. In this case statistics would be most unhelpful because the odd one is the correct one! The results shown are, incidentally, typical of those obtained when specimen misalignment is overlooked and 'rounding' at the yield point is obtained.

    Good practices in recording and checking data. Systematic recording of data is frequently overlooked but the experimenter must learn to overcome this. Tables prepared *before* the tests—columns of this and that will save time. The tables should never be recorded on scraps of paper and an exercise book for this purpose must be used. Enough room alongside the tables should be available for jotting down notes of observations made at particular times. Graph paper of various

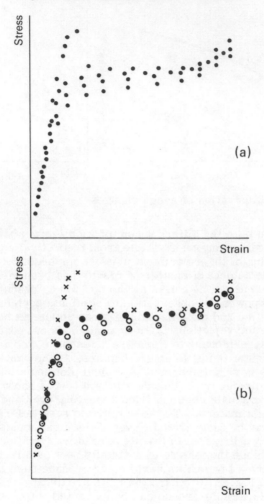

FIG. 2. 7   Scatter in tensile testing

sizes and types should be available for use *during* the experiment. Graphs and any other records should be attached to the exercise book and numbered and dated in an orderly fashion. At the time of the experiment no fact should go unrecorded. When the tests are completed and the results and conclusions are being communicated through a report all the information in the exercise book can be carefully sifted and correlated. Much time and many moments of frustration can be saved by careful organisation of effort along these lines.

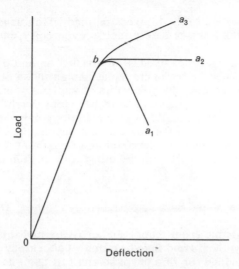

FIG. 2. 8   Load deflection characteristics of a compression loaded structure

So far as checking data is concerned some methods have already been mentioned. Other important ones are also available. Wherever possible simple checks of balance-energy, mass, momentum, current, forces and so on can be tried. Is a maximum or minimum expected? What is the significance of a measured asymptotic response? What, if anything, is obtained by extrapolating back to zero or to large values? Should a steady state be expected? These are all questions which the experimenter has to ask himself and for which a knowledge of the rudiments of the physical background of this problem is essential.

Tests of qualitative nature

There are some occasions when all that is needed of an experiment is a qualitative description of a design or phenomenon. Careful observation and measurement are still essential but the modelling of a real situation by a simplified test may be possible.

A good example of this approach occurs in structures required to carry loads in compression. The load-deflection characteristics of such structures may be of the forms shown in Fig. 2. 8. The most dangerous structures are those following the path 0a1 because these are liable to violent collapse once the bifurcation point b is reached. Those following paths 0a2 or 0a3 are safe even though they suffer large deformations after the bifurcation point. The type of behaviour is not easily calculated and is a function of the geometry, material and type of structure involved. Under these circumstances the type can best be ascertained by a preliminary model test to obtain a qualitative picture of the likely behaviour. If the path 0a1 is found then a new design for the structure can be sought or at the very least a high safety factor assigned to the bifurcation load which usually can be calculated. For the safer paths a lower safety factor, even unity, can be applied to the bifurcation load.

Another device sometimes used is to conduct an analogue test. Many physical phenomena are governed by the same equations as those of electrically conducting paper. Thus by measuring the potentials at various points in the paper, analogues of other situations and the importance of boundary conditions or shape changes can be assessed. Another example is in the analogy between the profil of soap bubbles and the torsion of bars. Such devices are beyond description in this book and are mentioned only in principle—the principle being that if it is easier to do one test rather than another and still gain meaningful data, then do it!

Summary

1. Experimentation is capable of and best approached by method. Discipline in this is vital.
2. Experimentation is inseparable from theory and analysis and this can take many roles as aids to Design, Synthesis and Investigation. Laboratory courses for student engineers are provided for this purpose and full use should be made of them.
3. Above all else, healthy scepticism of results, whether they are obtained by the experimenter or by others, should be generated. Motives should be questioned at every step before progressing to the next step. The next step should only be taken when satisfaction in these enquiries is ensured.

References

1. ARTHUR GORDON. *History of Flight*. American Heritage Publishing Company, New York, 1962.
2. Table 2.1 is a modified version of that given by W. G. Wood. 'A Methodology for Experimental Work'. *Bull. Mech. Eng. Educ.*, Vol. 7, 1967, pp. 47-9.

3 *Dimensional Analysis Applied to Experimentation*

3.1   The applications of dimensional analysis

Dimensional analysis is founded upon the observation that any functional relation that is an analytical model of a real event is independent of the system of units used in measuring that event. This analysis can have the following consequences for experimentation:

1. It can greatly reduce the amount of experimental investigation by reducing the number of independent variables.
2. The effect of one variable can be determined by an experimental variation of another.
3. The applicable range of a variable can be extended beyond the experimental range. In the extreme case this applicable range can be achieved by the measurement of only a single value of each variable.
4. It can show that sometimes a quantity has no effect upon the phenomena and so can be excluded as an experimental variable.
5. The oversight of an independent variable in the planning of an experiment can be revealed.
6. The cost of an experiment can be reduced or even sometimes experimentation can be made feasible by enabling tests to be made on reduced scale models of the full size system. Alternatively it can ease experimental difficulties by enabling experiments to be performed on larger scale models of very small systems.

In this chapter it is assumed that the methods of Dimensional Analysis in the form of the principle of similitude and of Buckingham's Π Theorem [1] are familiar to the reader. A very brief summary of this theorem is given in Appendix 1. Discussion here is limited to an explanation and a demonstration of the consequences just described together with others.

3.2   Superfluous variables

For an experiment, care must be taken in distinguishing between dependent and independent variables. An illustration of this is provided by the design of an experiment to determine the power necessary to drive a fan which is pumping air along a duct. This power, P, might reasonably be expected to be a function of the fan diameter d, the air density, ρ, the air velocity, V, and the fan rotational speed, n. Or, written in standard notation,

$$P = P(d, \rho, V, n) \tag{3.1}$$

But it might be thought that the power depends also upon the rise in pressure across the fan, Δp, so that,

$$P = P(d, \rho, V, n, \Delta p) \tag{3.2}$$

This is incorrect. The variable Δp is superfluous, for it can be regarded as an alternative dependent variable so that

$$\Delta p = \Delta p(d, \rho, V, n) \tag{3.3}$$

and experiment would confirm this. Thus in equation (3.1) if d, ρ, V, and n are specified in value so that from equation (3.3) Δp is fixed then in equation (3.2) Δp is superfluous as it is not a variable that can be varied in complete independence of all the others.

Sometimes the exclusion of such superfluous variables is more obvious. For instance the fan thrust, T, must not be included as an independent variable because there is the analytical relation

$$P \propto ndT \tag{3.4}$$

and so again if d, ρ, V, n are specified then, from equation (3.1), P is fixed and so from equation (3.4) T is determined and is thus not completely independent.

3.3   Avoidance of superfluous experimentation

The application of dimensional analysis reduces equation (3.1) to a functional relationship between just two variables, each being a non-dimensional group. This relation can be formed as

$$\frac{P}{\rho n^3 d^5} = f\left(\frac{V}{nd}\right) \tag{3.5}$$

If an experiment is designed on the basis of equation (3.1), then a variation of only d would give a single curve on a graph of P plotted against d such as is sketched in Fig. 3.1(a). The additional variation of ρ would result in a family of curves as sketched in Fig. 3.1(b); the variation of V would result in a set of such graphs as shown in Fig. 3.1(c); and finally the variation of n would result in a family of sets of such graphs as seen in Fig. 3.1(d). The result is that both the experiment and the results are cumbersome in their extent. But if instead the design of the

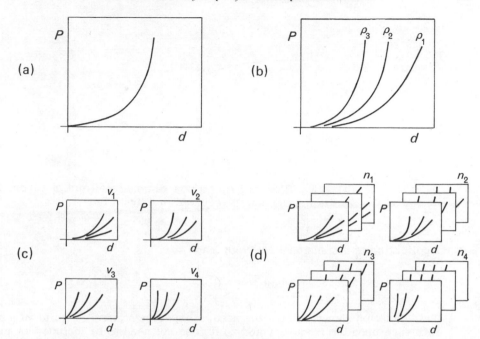

FIG. 3. 1   Illustration of a dimensional plot of experimental results
for the power required to drive a fan

experiment is based upon equation (3. 5) in which, in effect, the five variables of
equation (3. 1) have been reduced to two, three quite outstanding benefits ensue:

(a) only one independent variable, such as for example n, need be
varied in the experiment because this results in the required variation in the group V/nd.

(b) all the results plot as a single curve on a single graph such as
is sketched in Fig. 3. 2.

(c) the numerical results are independent of the system of units
used in the experimental measurements because the two groups of
equation (3. 5) are non-dimensional.

An extreme example is given by the oscillation of a solid in relation
to a point to which it is attached by a spring. If the spring characteristic is a
linear one, so that the extension force is proportional to the extension through a
spring coefficient of proportionality, e, then the frequency of oscillation, ω, might
reasonably be assumed to be a function of e, the mass of the object, m, and the
amplitude of the oscillation, a; or,

$$\omega = \omega(e, m, a) \tag{3. 6}$$

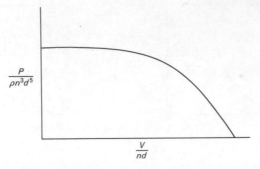

FIG. 3. 2   Illustration of a non-dimensional plot of experimental results for the power required to drive a fan

Application of dimensional analysis leads to

$$\frac{\omega^2 m}{e} = \text{constant} \tag{3.7}$$

This equation shows, without the need of experiment, that the value of a has no influence upon the frequency and so it does not need to be included as an experimental variable. Furthermore, dimensional analysis has revealed the nature of the functional dependence of ω upon both m and e. All that is required of an experiment is a test at a single value of each of ω, m and e so that the constant in equation (3. 7) might be determined.

The example just given is one for which the analytical solution is easy. But a similar application of dimensional analysis can be made for the vibration of a complex structure. Replacing the spring coefficient by E, the Young's modulus for the material of the structure, and replacing the object mass by the density of the material, ρ, and introducing a scale length, l, equation (3. 7) becomes,

$$\frac{\omega^2 \rho \, l^2}{E} = \text{constant} \tag{3.8}$$

The discussion of the usefulness of equation (3. 7) applies to equation (3. 8), which is valid for a situation where accurate analysis would not be possible.*

———————————————

* Equation (3. 8) might not always be adequate. For instance, in a complex structure non-linear spring effects may be present and the amplitude a is of significance and so the group a/l would be added to equation (3. 8) as a further independent variable.

3.4   Number of independent groups

To return to the previous example of the flow past a fan, the fan efficiency, η, might also be written

$$\eta = \eta(\rho, n, V, d) \tag{3.9}$$

If in addition the fan shape, as distinct from its size, is varied by altering the angle, θ, of the blades to the plane of the fan*, then,

$$\eta = \eta(\rho, n, V, d, \theta)$$

or using dimensional analysis,

$$\eta = f\left(\frac{V}{nd}, \theta\right) \tag{3.10}$$

As with a previous example, dimensional analysis has enabled one variable, in this case ρ, to be excluded.

Equation (3.10) might have the form sketched in Fig. 3.3. If only the peak values of the efficiency, η_P, are of interest then a relation between θ and V/nd

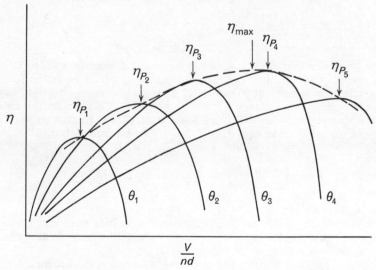

FIG. 3.3   Illustration of a non-dimensional plot of the efficiency of a fan

---

* It is usual to think of this as a change of pitch which is equal to $\pi d \tan \theta$. This configuration is called a variable pitch fan.

is implied as illustrated in this figure. Thus equation (3.10) reduces to the alternative forms,

$$\eta_P = f\left(\frac{V}{nd}\right) \tag{3.11}$$

and,

$$\eta_P = f(\theta)$$

the former being shown as the dotted curve in Fig. 3.2. Further, when only the maximum value of the efficiency is to be determined then equation (3.11) reduces further to

$$\eta_{max} = \text{constant}$$

This search for certain values, such as maxima, minima and zeros is a common experimental requirement and, as shown, it reduces the number of non-dimensional groups.

Returning to equation (3.9), it reduces to

$$\eta = f\left(\frac{V}{nd}\right) \tag{3.12}$$

and as before mentioned only one of V, n, and d need be varied in an experiment. But supposing two of these variables were varied. If equation (3.9) was valid a single curve of η against V/nd would still be obtained. It might happen that the results did not collapse into such a single curve. This would indicate that there is an independent variable missing from equation (3.9). In this case it would be the viscosity of air, μ, so that equation (3.12) would be corrected to

$$\eta = f\left(\frac{V}{nd}, \frac{\rho V d}{\mu}\right)$$

The combination of the use of dimensional analysis with an extension experiment in which one extra variable is changed provides this most useful check against the omission of a significant variable.

However, this check might still not reveal a missing variable. For if such a variable had a value of zero or infinity in the experiment then the missing group containing it would have the constant value of zero or infinity. An adjustment of another variable would not show the effect of this missing group.

Sometimes use can be made of a relation basic to the phenomena being investigated. For instance, if a structure fails under elastic instability then the failure load F, can be expressed in terms of Young's modulus E, and a scaling factor l, or

$$F = F(E, l) \qquad (3.13)$$

If the structure is a single strut and the separate effects of the strut length, l, and the cross-section size w, are to be determined, then equation (3.13) extends to

$$F = F(E, l, w) \qquad (3.14)$$

which becomes*

$$\frac{F}{El^2} = f\left(\frac{w}{l}\right) \qquad (3.15)$$

But we know that in bending E appears in the combination EI where I is a representative second moment of area of the strut section.
Also

$$I \propto w^4$$

so that equation (3.14) becomes

$$F = F[(EI), l]$$

giving

$$\frac{Fl^2}{EI} = \text{constant} \qquad (3.16)$$

where the constant will be a function of the end conditions. Thus the extra information about a fundamental feature of bending results in a reduction of the number of groups from two in equation (3.15) to one in equation (3.16). Experimentation is then only required to obtain the value of the constant in the latter equation.

A different example of the use of extra information is provided by the application of dimensional analysis to the motion of the planets. Suppose the time of orbit, t, is a function of the mass of the planet, m, the gravitational force exerted on the planet by the sun, F, and a size of the orbit, l. Then

$$t = t(F, m, l)$$

---

* Two groups are obtained because only two dimensions are required, that is, L and the combination (MT^{-2}).

or

$$\frac{t^2 F}{ml} = \text{constant} \qquad (3.17)$$

But extra information is available in the form of Newton's law of gravitation which says that

$$F \propto \frac{mm_s}{l^2}$$

where m_s is the mass of the sun and is a constant. Insertion of this relation into equation (3.17) gives

$$\frac{t^2}{l^3} = \text{constant}$$

This is Kepler's famous law which he derived from extensive observation. Use of the extra information has here enabled a different formulation of the non-dimensional group of equation (3.17) to be made.*

3.5   Effectiveness of experimental variables

Returning to the example of the efficiency of a fan, equation (3.10) shows that if interest is limited to a fan of a single shape so that θ is fixed in value then

$$\eta = f\left(\frac{V}{nd}\right) \qquad (3.18)$$

This reveals another marked usefulness of dimensional analysis because, in an experiment to determine the form of this function, only one of V, n and d need be varied. The effect of variation of the other two is then determined without the need for further experimental change of them. Obviously some variables are more conveniently changed than others. In this example change in d would involve the manufacture of a family of fans, and change in n would require a variable speed drive motor which, certainly in large sizes, would be costly in comparison with a fixed speed one. In contrast, V might more readily be varied over all the finite range illustrated in the sketch of Fig. 3.3 by an adjustable throttling of the flow through the containing duct.

The choice of experimental variable is not always an arbitrary one. Suppose an experiment is planned to determine the gradient of pressure, P, along a

---

* This example illustrates a way of using Dimensional Analysis; it does not correspond with the historical development of these relations.

straight, circular pipe along which a gas of density ρ, and viscosity μ is flowing with a mean velocity V. As well as ρ, μ and V the independent variables that affect P will include the height of the surface roughness on the pipe, ϵ, and the pipe diameter, d. Thus,

$$P = P(\rho, \mu, V, \epsilon, d)$$

The three corresponding non-dimensional groups can be written

$$\frac{Pd}{\rho V^2} = f\left(\frac{P\rho\epsilon^3}{\mu^2}, \frac{\rho V d}{\mu}\right) \tag{3.19}$$

Now experiments would be expected to plot as a family of curves.

If the most easily adjusted variables in this gas flow are V and ρ then P could be measured as a function of them. For instance, it might be planned to vary P and ρ, keeping $(P\rho)$ constant so that a plot is obtained between the first and third groups of equation (3.19) as a single curve with the second group a constant. If an attempt was made to obtain a family of curves by repeating this experiment for other values of $(P\rho)$ it would be found that the original single curve was always repeated. A tendency would be to conclude that the second group of equation (3.19) did not affect this phenomenon; in this case such a conclusion would be wrong. The reason can be seen by reformulating equation (3.19) in the form

$$\frac{Pd}{\rho V^2} = f\left(\frac{\epsilon}{d}, \frac{\rho V d}{\mu}\right) \tag{3.20}$$

which shows that the proposed experiment, in which ϵ/d is always held constant, would indeed give only a single curve. A check on the use of equation (3.19) to plan an experiment can thus be made as follows. The variables that it is planned to hold constant are d, μ and ϵ: inspection of them shows that they can be used to form a non-dimensional group, ϵ/d, which is the one appearing in equation (3.20). This form of check should always be made.

3.6 Investigation by using scale models

A full-scale test of a system can often be impracticable and then dimensional analysis shows the way to testing on scale models.

Suppose it is required to know the steady state voltage distribution in a liquid of low conductivity, the system of interest might be an oil insulated transformer. Then the voltage ϕ could be regarded [2] as a function of a boundary reference voltage, ϕ_0, the coordinate, r, a size of the system, l, the conductivity, λ, the dielectric coefficient ϵ, and the coefficient of diffusion of the conducting ions, D. These seven variables form into

$$\frac{\phi}{\phi_0} = f\left(\frac{D\epsilon}{\lambda l^2}, \frac{r}{l}\right) \tag{3.21}$$

If, in a model test, the independent groups, $(D\epsilon/\lambda l^2)$ and r/l, each have the same numerical values as the full-scale ones then, whatever the form of the function of equation (3.21), the dependent group, ϕ/ϕ_0 must also have the same numerical value at model as at full scale: the form of the function of equation (3.21) does not have to be determined when setting up a test at model scale.

     Constancy of the group $(D\epsilon/\lambda l^2)$ means that tests on a reduced size model can only be made if the electrical characteristics of the liquid are changed. It is possible that λ alone could be increased to reduce l in accordance with

$$l^2 \propto \frac{1}{\lambda} \tag{3.22}$$

For example, a hundredfold increase in λ results in a one-tenth scale model. Equation (3.21) also shows that a reduced boundary voltage ϕ_0 can be used in the model test, constancy of the group ϕ/ϕ_0 resulting in ϕ values being correspondingly scaled down.

     The current density, j, through this type of conductor varies with time, t, after the initial application of the voltage ϕ_0 so that

$$j = j(\phi_0, l, \lambda, \epsilon, D, t)$$

This reduces to

$$\frac{jl}{\lambda\phi_0} = f\left(\frac{D\epsilon}{\lambda l^2} \frac{\lambda t}{\epsilon}\right)$$

Now the third group shows that in a model test the time can be scaled by changing the conductivity alone in accordance with

$$t \propto \frac{1}{\lambda} \tag{3.23}$$

Then the first group shows that the current density can be adjusted by changing one or more of l, λ and ϕ_0 to meet the requirement that

$$j \propto \frac{\lambda\phi_0}{l} \tag{3.24}$$

To provide a numerical example for a model conductivity 100 times that of full scale then from equation (3.22) the model would be one-tenth of full size.

     Equation (3.23) shows that a time scale in the model test would be one hundredth of the full-size scale. The applied electrical field will be proportional to ϕ_0/l. If it is desired to keep this the same in the model test as at full scale then the applied voltage in the model test, ϕ_0 will be one-tenth of the full-scale voltage. It then follows from equation (3.24) that the model current density will be 100 times that of the full-scale value. Finally the leakage current, which will be proportional to jl^2, will have the same value in the model test as at full scale. Thus dimensional analysis shows how all the variables are scaled in the model test.

Model tests may be used for a surprising number of phenomena. For example, the problem of thermal effects in solids is often soluble by model tests by analogous arguments to those just given for the electrical phenomena [3]. Yet surprisingly the many possibilities are little appreciated. The following is an example of a slowly growing awareness [4].

'I learnt the hard way about Gipsy Moth's sailing characteristics and so was much interested in Hobbies sailing model of Gipsy Moth IV.

'When Hobbies made their first prototype hull to scale, it had "all the appearance of static stability" when the first flotation tests were carried out. When the sails were set and sailing trials began it was found, however, that the keel was not heavy enough, and that the boat tended to heel over on its side, immersing the sails in the water. This is very similar to what occurred when Gipsy Moth IV underwent her first sailing trials. The keel was then redesigned and carried aft to the rudder post, just as was prescribed for Gipsy Moth in Sydney.

'It was also found that the mizzen sail "tended to tack over and veer the model off course", meaning that the mizzen sail was not balanced by the headsails, and consequently brought the model up into the wind. This same fault in Gipsy Moth IV itself was partly over-come in Sydney, when the topmast stays were moved farther forward. This balanced the boat on most points of sailing. In the model, the designers reduced the mizzen sail area, which produced the same effect.

'Finally, it was found that the speed and performance of the model were far beyond the makers' expectations. They thought that the hull design and rigging were now the best on the market, and then the model's performance was compared with full radio control racing yachts, it left those behind for both speed and controllability.

'My comment on all this is: What a pity that the designers of Gipsy Moth IV did not have time to make a model to sail in the Round Pond before the boat was built! What an immense amount of trouble, worry and effort this would have saved me, by discovering Gipsy Moth's vicious faults and curing them before the voyage!'

References

1. R.C. PANKHURST. *Dimensional Analysis and Scale Factors*, Chapman and Hall, London, 1964.

2. J.C.GIBBINGS. 'Non-dimensional groups describing electrostatic charging in moving fluids', *Electrochimica Acta*, Vol. 12, p. 106, 1967.
3. A.J.SOBEY. 'Advantages and limitations of models', *J.R.Ae.Soc.* Vol. 63, No. 587, November 1959.
4. F.CHICHESTER. *Gipsy Moth Circles the World*, Ch. 8, p. 106, Hodder and Stoughton, 1967.

4 *Instrumentation*

4.1   Attitudes towards instrumentation

The importance of developing and adopting the correct attitude to instrumentation should be emphasised at the outset. To the engineer, instrumentation is a means to an end, and one should not be overawed by apparent complexity of equipment. It is not always necessary nor advantageous to have a full understanding of the various 'black boxes' employed, although clearly an appreciation of the basic principles is desirable. Equipment should be treated with respect and indiscriminate button pressing should be avoided. Manuals are normally available even on the more basic items, and these should be consulted if necessary.

As a general rule, an initially untrusting attitude towards equipment should be developed, until it is shown from calibration tests and use that this is unwarranted: trust towards another individual comes only with time and experience, and so it is with instruments. Following on from this comes the need for strict honesty in making instrument observations. This should go without saying, but it is a natural tendency to want to ignore faulty equipment or apparently inconsistent results when time is limited. Should an instrument be suspected of malfunctioning, it should be adjusted or replaced, or if this is not possible, the likely error involved should be acknowledged. Admission of limitations adds credence to reported experimental results.

4.2   Equipment selection

The basic function of engineering instrumentation is to convert physical quantities, which are not normally themselves directly measureable, into quantities which are. Invariably the ultimate quantity employed is linear (or angular) displacement. For example:

U-tube manometer : change in pressure—change in fluid level.
Thermocouple : change in temperature—change in e.m.f.—needle rotation on voltmeter.

Spring balance : change in load—deflection of spring.
Pressure transducer : change in pressure—deflection of diaphragm—
change of strain—change in e.m.f.—scale
change on recording equipment.

There are many factors involved in the choice of equipment, not least of which are *range* and *accuracy*. Clearly, the two are related in that accuracy will be reduced if the portion of the total instrument range used is small. Normally the range to be measured will be known within reasonable bounds, and a sensible choice of instrument can be made. For example, whether to use a U-tube manometer or a Bourdon tube type pressure gauge; whether to apply load by means of dead-weights or using a tensile testing machine; whether to measure dimension changes with a micrometer or a metre rule, etc. *Accuracy* should not be confused with *precision*: accuracy is concerned with absolute correctness of measurement, whereas precision entails consistency, i.e. repeatability and stability with time (see Ch. 5, sect. 5.1, 'Accuracy and precision').

Other factors encountered in the choice of instruments are:
Sensitivity, defined as the ratio
$$\frac{\text{index movement}}{\text{corresponding change in measured quantity}}$$
2. *Discrimination* (or Resolution), defined as the smallest change in the measured quantity that can be detected. This assumes importance in 'null-balance' methods, examples of which are Wheatstone bridge techniques and micromanometers.
3. *Dynamic response*, which is to do with the ability of an instrument to respond accurately to a rapidly changing input. Systems where this assumes importance are, for example, the U-tube manometer, moving-coil galvanometers and various other more complex recording instruments. The response of such instruments can be approximated by the single degree of freedom, second order, oscillating

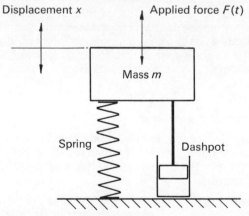

FIG. 4.1   Oscillating mass system

spring system shown schematically in Fig. 4.1. and represented by the differential equation,

$$m\frac{d^2x}{dt^2} + c\frac{dx}{dt} + kx = \text{Applied force } F(t) \tag{4.1}$$

where m is the mass of the body, c the damping coefficient of the dashpot and k the spring stiffness. The following analogies can be drawn:

Instrument inertia—inertia force $\left(m\dfrac{d^2x}{dt^2}\right)$

Instrument resistance—damping force $\left(c\dfrac{dx}{dt}\right)$

Instrument stiffness—spring force (kx)

Instrument signal—exciting force $F(t)$

Instrument deflection—displacement (x)

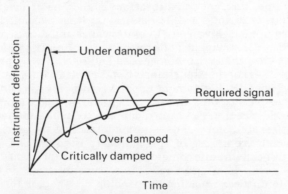

FIG. 4.2   Damping characteristics

Figure 4.2 illustrates the categories of response that can be obtained when the instrument signal is applied instantaneously and then maintained constant. For situations where damping can be controlled it is common to have slight under-damping which tends to improve reading accuracy.

4. *Interference* with the experimental set-up. Examples of this would be the reinforcing effect of a strain gauge on a thin plate test specimen, the effect of a flow measuring device inserted into a pipe line, or lead resistance effects in electrical circuits.

  All of the above factors may have to be considered in selecting the correct instrumentation for the job to be done. Of course, the student will often find that the choice of equipment for his laboratory exercise has already been made, or that in the case of projects the most suitable instrument is not available. This is often unavoidable, but an appreciation of the above factors will assist in the assessment of results, and constructive criticism of experimental apparatus is always acceptable in laboratory reports. In any case it is useful, at an early stage, to learn to improvise when the quality of available equipment is limited. This is invariably the case when real problems are tackled with limited time and finances.

4.3   Calibration

Calibration is the comparison between one instrument and another known to be more accurate, under conditions as nearly as possible identical to those existing in the test set-up. The basic standards for the six fundamental quantities, length, mass, time, electric current, temperature and luminous intensity are controlled, in so far as Great Britain is concerned, at the National Physical Laboratories, Middlesex. From these basic standards, local standards are produced from which calibrations can be conducted as required. Clearly, the greater the number of calibration 'steps' from the basic standard, the less reliable will be the instrument in question.

Strictly speaking, check calibrations should be conducted before every test series, but where this is not possible preliminary runs should be conducted to give a 'feel' for the equipment involved. Due allowance must be made for possible inaccuracies. Records of instrument calibrations are often filed as a matter of routine and, as a first step, these can be checked and used if they are up to date.

Sources of error in experimentation are dealt with in detail in the following chapter, and specific instrument errors commonly encountered are included later in the present chapter. However, in general, apart from the more obvious sources like incorrect zero setting, damage to indicating mechanisms, etc., there are a number of contributory causes to variations of measuring instruments with time, amongst which are ageing of springs, wear in joints and friction effects. Examples of these can occur in micrometers, dial gauges, pressure gauges and stop-clocks. The net result is often to cause a hysteresis effect in the instrument output which results in different readings depending on whether the required value is approached from below or above.

The importance of identical calibration and operating conditions has already been emphasised. One example illustrating this is the misuse of dial gauges for deflection measurement. These gauges are calibrated to measure movement in the axial direction of the spindle and, clearly, if the dial gauge is not set up with the spindle in line with the required movement, then errors, usually termed 'cosine errors', will be introduced. For example, a 10° misalignment will give an error of 0·15 mm in 10 mm. There is little point in using an instrument capable of reading to within 0·01 mm if the correct operating conditions are not met.

Where more sophisticated electronic equipment is involved, errors are often introduced by not complying with the specified 'heating-up' period. This can in some cases be of the order of say 10 minutes during which time apparent results could vary considerably. Another important cause of errors in electrical equipment is that of interference effects from adjacent equipment. Techniques for dealing with this include careful earthing and the use of screened leads, but other-wise the problem is beyond the scope of this text.

4.4   Summary

The foregoing can be summarised in three prerequisites in the use of equipment and instrumentation, if the best results are to be obtained:

(a)   the need to develop an initially untrusting but respectful attitude to instrumentation at large;

(b)   the importance of correct equipment selection for the job on hand;

(c)   the need to ensure through careful calibration that the results obtained are meaningful.

4.5   A compendium of engineering laboratory equipment

In the following sections an attempt is made to outline the various categories of instrumentation that are likely to be encountered in an undergraduate engineering course. The principles of operation are briefly explained and accuracy and range of operation discussed. It is not claimed that the list is complete; the variety of equipment and instrumentation available nowadays precludes a fully comprehensive survey, and where further details are required reference can be made to one of the many excellent text-books available on the subject, some of which are listed at the end of this chapter.

Measurement of length

Strictly speaking, the measurement of length is only one aspect of the general field of mensuration which includes measurement of length, angle, area and volume, but since the latter three are derived from length measurements we will be concerned here mainly with length. The field of Metrology, i.e. the application of mensuration methods to workshop practice, is not covered in detail, this being a study in its own right, e.g. Hume [6], although mention is made of the more basic workshop tools.

For general dimensioning of components *wood or steel rules* are normally satisfactory. Wood has a smaller thermal coefficient of expansion than steel, which can be an advantage, but on the other hand steel rules are more readily engraved for greater precision and are not so susceptible to end wear. Nominal dimensions, e.g. tube and shaft diameters, are normally measured by means of machinists' *calipers* used in conjunction with a steel rule. Surprising accuracy can be achieved with this relatively crude method. For example, it is quite possible to measure the diameter of a shaft, consistently to within ±0·2 mm of its true dia-meter, in this way.

For more accurate dimensioning, use is made of a *vernier caliper* or a vernier height gauge. By means of the vernier scale an additional significant figure in the measurement can be obtained over conventional rules.

Suppose a dimension is required to the nearest 0·1 mm using vernier calipers. The vernier and main scales would be as shown in Fig. 4.3. The ten divisions of the vernier scale are spaced over 0·9 cm of the main scale and the aim is to select the vernier graduation which is nearest to being in line with a main scale graduation. In the illustration, this appears to be the digit 3 and the required dimension is thus 20·3 mm.

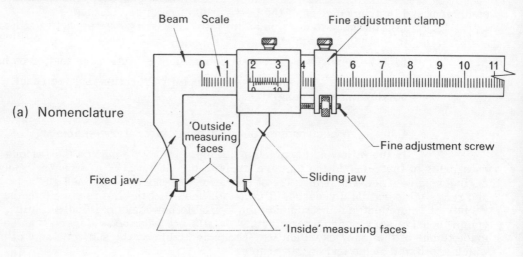

(a) Nomenclature

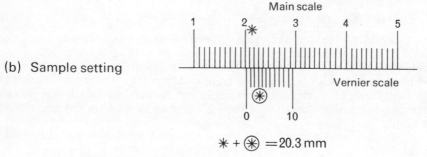

(b) Sample setting

$$* + ⊛ = 20.3\,\text{mm}$$

FIG. 4.3   Vernier calipers

The micrometer caliper or simply *micrometer* is, as the name suggests, a direct-reading caliper arrangement, the constituent parts of which are shown in Fig. 4.4. Accuracy of measurement is superior to the vernier caliper and is obtained by means of a finely threaded screw, e.g. 16 threads per cm, to which the graduated 'thimble' is attached. The required reading is a combination of the main scale and the thimble scale. Increased precision is obtained by increasing the thimble diameter to say 60 mm thus giving an enlarged scale, and in some cases a circular vernier scale is incorporated. In the inset to Fig. 4.4 a sample

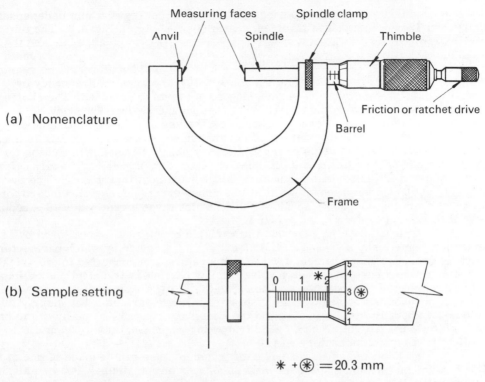

(a) Nomenclature

(b) Sample setting

$* + \circledast = 20.3$ mm

FIG. 4.4   25 mm Micrometer

setting is given and the reader should convince himself that the required dimension is again 20·3 mm. If in good condition, a standard 25 mm micrometer should be capable of reading to within 0·005 mm. Typical sources of error are zero error, wear and tear, dust particles between mating faces, differential temperature effects and reading errors due to inexperience or carelessness.

The *toolmaker's microscope* is a useful tool which incorporates two micrometer heads attached to a floating surface table on which the component to be measured is located. Sizes can thus be obtained in two dimensions and the microscope attachment ensures consistent alignment, apart from the increased accuracy due to enlargement of the component.

Mention should be made of the use of *slip gauges* in the calibration of micrometers and vernier scales. These are small steel blocks with two highly finished parallel surfaces such that when two blocks are brought together with a sliding action they will stick together due to molecular attraction. The individual slip gauges are of varying thickness, so chosen that, for example, from a set of 87 blocks, any dimension in the range 0·500 mm to 100·000 mm can be obtained to within an accuracy of 0·001 mm. Micrometers and verniers can be checked for accuracy against a range of slip gauge stacks, the accuracy of which is dependent on the number of slip gauges involved.

Measurement of time and speed

Measurement of time becomes necessary when the frequency or speed (linear or rotational) of engineering components is required. The choice of equipment will depend on the experimental set-up and desired accuracy.

For many laboratory exercises, the use of manually operated *stop-clocks* will give sufficient accuracy, although, since a ratchet mechanism is usually involved giving discrete steps of say 1/5 s, they should be used with caution for short periods of time. Frequent checking of laboratory stopclocks, ideally before every experiment, is advisable since the rough treatment to which they are inevitably subjected has been known to cause gross errors. A useful standard for calibration is the telephone 'speaking clock' which gives a continuous 10 s repeating signal to a precision of around ± 1/10 s.

It is often the case that the variable of interest is monitored by means of some form of *transducer,* Neubert [7] i.e. a device by which parameters such as pressure, displacement, torque, temperature, are converted to electrical outputs, in which case, the times involved in any particular transient can be measured in a number of different ways. The *cathode ray oscilloscope* (c.r.o.) is a useful general purpose instrument for the display of a single variable. The display tube is usually about 10 × 8 cm in size and the time base range is normally of the order of 1 s/cm to 0·5 μs/cm. It has a wide frequency response, ranging from d.c. to 3 MHz for a typical laboratory unit.

For permanent recording of transients use can be made of one of the large range of variable speed *graphical recorders* now available. Pen or galvanometer recorders can be used for frequencies up to about 100 Hz. For frequencies in excess of this, the *ultra-violet* recorder is particularly suitable. These are normally multi-channel instruments, the output being displayed on photosensitive charts of the order of 15-30 cm in width. Chart speeds vary typically within the range 1 mm/s to 2 m/s. With a suitable choice of galvanometer element, frequencies of up to 5 KHz can be recorded. Transients can also be recorded permanently by means of *trace-recording cameras* used in conjunction with cathode ray oscilloscopes. The potential advantages of this system over graphical recorders are the higher range of frequencies, and the fact that transients can be examined prior to recording which can reduce the cost of recording material. *Storage oscilloscopes* have a facility for retaining the display semi-permanently, and in some cases a 'split-screen' facility is available for comparing the display with some standard trace.

In the field of vibrations, *frequency counters* are often employed and are capable of time interval measurement in the range of 1 μs-10^4s (2·8 h). They rely for their accuracy on the maintenance of a set frequency—usually in the form of ultra-stable internal oscillators, and with the aid of suitable transducers, frequencies in the range 10 Hz-1·2 MHz can be monitored directly by a typical laboratory instrument.

In the measurement of rotational speed, the simplest method is to count the revolutions over a period of time by means of a stopwatch, together with a *revolution counter* connected to the rotating component. Alternatively a mechanically operated *tachometer* graduated to read directly in revolutions per minute may be used. Electric tachometers are also available which generate an output voltage proportional to speed, and with a suitable choice of scale, can again read directly in revolutions per minute.

It is often the case that oscillatory or rotational speeds are required without interference from mechanically attached transducers. A common technique here is to use a *stroboscope*. This is basically an inert gas-discharge lamp which can be flashed intermittently at continuously variable frequencies. When the frequency of the illumination coincides with the frequency of motion, the moving part will appear to be stationary.

More recent developments in this sphere are electrically operated tachometers and counters. These can involve electro-magnetic, capacitive or photo-electric probes. To take the first as an example: irregularities in the mechanism such as gear teeth, keyways, etc., produce a momentary change in the circuit reluctance as they pass the probe thus inducing an e.m.f. in the transducer coil, the frequency of which is proportional to the frequency of the motion, or the shaft speed as the case may be. The output from these transducers can be arranged as a digital display using the frequency counter mentioned above.

Measurement of displacement

(i) *Mechanical methods*. Much of what has been said on the subject of length measurement will apply equally well to the measurement of displacement. For example, the increase in length of a section of wire under load could be measured satisfactorily by means of vernier scales. Differences in manometer fluid levels can be measured accurately using a micrometer head. (See 'measurement of pressure', p. 53). However, for the measurement of static or quasi-static displacements use is normally made of *dial gauges* which can be adapted to suit most circumstances, and give moderate precision. The basic instrument is shown in Fig. 4. 5, the useful range in this case being 0-10 mm. The 50 mm diameter main dial is graduated in 100 divisions of 0·01 mm and the small dial in 10 divisions of 1 mm, each corresponding to one complete revolution of the main dial. The principle of operation is a rack and pinion mechanism with a spring to keep the measuring tip in contact with the work piece. The main dial may be manually rotated to facilitate zero adjustment. Calibration is readily checked using a micrometer jig, or slip gauges.

For the measurement of strain, e.g. in a tensile test, use is sometimes made of *mechanical extensometers*. There have been many variations in the design of these over the years, the most common being the Huggenberger and Johansson types. They rely on mechanical methods for the magnification of the change in the gauge length and have the advantage that they can be re-set to extend the useful range. However, on account of their physical size, cost and the need for individual reading, they are normally unsuitable for use in large numbers, in inaccessible

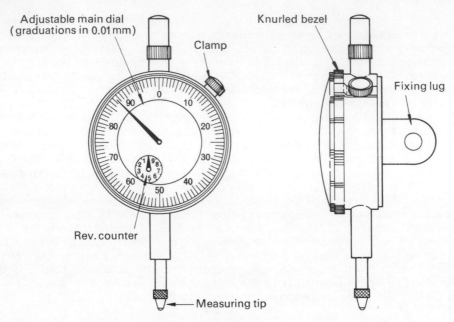

FIG. 4. 5   10 mm dial gauge

locations or in hostile environments. Mechanical extensometers have been largely superseded by electrically operated displacement transducers as discussed in (iii) below.

(ii) *Optical methods*.   Rotational displacement, in particular, is conveniently measured by optical methods. The advantages of using light as a means of measurement are that, (a) it travels in a straight line and can act as a long lever arm to magnify rotation and, (b) the mass of light can be taken as zero and hence there is no interference with the moving part.

A typical set-up is shown in Fig. 4. 6 using a *theodolite* (or a telescope fitted with cross-hairs), a small mirror and a graduated scale. The theodolite is placed at a distance l from the structure and sighted on the mirror attached to the specimen such that the scale is seen in the field of view. If the reading from the cross-hairs on the scale changes by Δs, then the angle of rotation of that part of the specimen to which the mirror is attached will be $\frac{(\Delta s)}{(2l)}$ radians. This basic principle can be adapted for use in a variety of situations. For example, with a little ingenuity it is possibe, with the use of simple mechanisms, to convert linear displacement to rotational, and thus use the same optical technique.

For more accurate measurement of rotational displacement the *autocollimator* principle is often used. This is shown schematically in Fig. 4. 7 (Evans and Taylorson) [5]. It consists of a micrometer microscope A, an illuminate

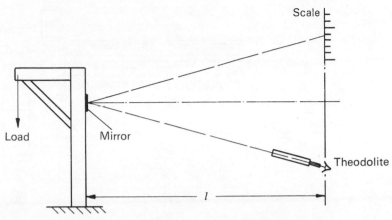

FIG. 4. 6   Measurement of rotational displacement

cross-wire B, a collimating lens C situated such that B is in the focal plane, and a high-quality reflecting surface D. As illustrated, an angular movement θ of the reflecting surface results in a movement of the cross-wire image by a distance d, which can be measured by the micrometer microscope. The required angular rotation will then be given by $\theta = \dfrac{(d)}{(2f)}$ radians.

(iii) *Electrical methods.*   There is a variety of electrical methods now available involving *displacement transducers* of one kind or another. These have the advantage over mechanical methods in that they can be used for dynamic measurements. One common type employs varying inductance as the output to be recorded. This is produced by the movement of a permeable core in the form of a rod of the order of 3 mm diameter inside a double coil arrangement as shown in Fig. 4. 8. The coils are wound in such a way that the impedance-core displacement relationship is linear, the two coils forming two arms of a variable inductance bridge circuit. Transducers of this type are supplied over the range 0-3·0 mm to 0-250·0 mm.

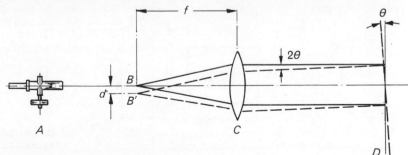

FIG. 4. 7   Autocollimator principle

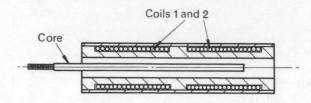

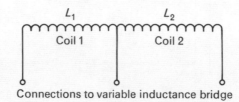

FIG. 4. 8   Variable inductance displacement transducer

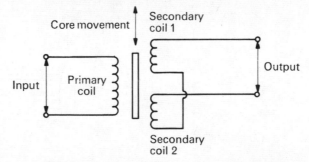

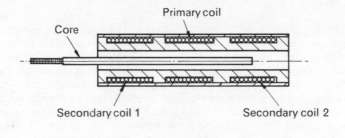

FIG. 4. 9   Differential transformer displacement transducer

The linear variable differential transformer (l.v.d.t.) is another common displacement transducer constructed in a similar way to the inductive type transducer, although with a different operating principle. In this case (Fig. 4.9) there are three coils involved. When the centre primary coil is energised with an alternating voltage supply, voltages are induced in the two secondary coils which are connected in such a way that the differential voltage output varies linearly with the displacement of the core. In both the above categories of displacement transducer the resolution is theoretically infinite, the effective resolution depending on the associated recording equipment.

Another category of displacement transducer relies on change of capacitance between two or more moveable plates, insulated from each other. These can be extremely sensitive devices but because of difficulties in associated instrumentation are not in common use and will not be discussed further here (Cook and Rabinowicz [3] give further details).

Probably the most common and certainly the most adaptable tool for use in the measurement of displacement is the *electrical resistance strain gauge*, and this being the case, it will be considered here in some detail. Emphasis will be given here to its transducer applications rather than to its use in the field of stress analysis. In principle, when a length l of wire is pulled in tension, l increases and the cross section A decreases thus increasing the resistance $\frac{(\rho l)}{(A)}$ of the wire, ρ being its resistivity. The strain gauge is virtually a length of wire made up in a grid form on a suitable nonconducting backing material. When the gauge is bonded to a test piece, in our case the transducer, it will be forced to strain either in tension or compression as the transducer is deformed by movement of the structure under test.

Although wire strain gauges are common, the grid is more often constructed by an accurate etching process, from foil material of the order of 0·004 mm thick. A typical foil gauge geometry is shown in Fig. 4.10. Gauge lengths

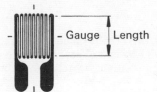

FIG. 4.10   Foil strain gauge

vary in the range $\frac{1}{64}$ in (0·4 mm)-4 in (100 mm), the most common sizes being $\frac{1}{8}$ in (3 mm) and $\frac{1}{4}$ in (6 mm). Provided all necessary precautions are taken and the gauge is properly bonded, mechanical strain can be measured with a precision as high as $\pm 1 \times 10^{-6}$.

Change in resistance of the strain gauge is usually measured by use of equipment based on the Wheatstone bridge circuit; this is shown in its basic form in Fig. 4.11. R_1 is the transducer strain gauge and R_2, R_3 and R_4 are either additional strain gauges or stable resistors depending on the technique employed. The bridge can be used as a null-balance system in which case the change in resistance of the active gauge R_1, causing unbalance of the bridge, can be obtained by adjusting the resistance R_4 by ΔR_4 such that balance is restored (i.e. $\Delta V = 0$).

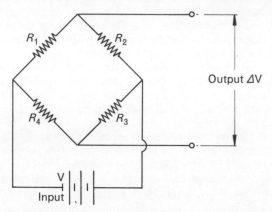

FIG. 4.11   Basic Wheatstone bridge circuit

From basic Wheatstone bridge theory we have the 'ohmic' strain

$$\frac{\Delta R_1}{R_1} = \frac{\Delta R_4}{R_4}.$$

The mechanical strain is given by

$$\frac{\Delta R_1/R_1}{K}$$

where K is the Gauge Factor provided by the strain gauge manufacturer. However, in transducer applications, there is no need to use K since we are interested only in obtaining an output proportional to displacement, which can be subsequently calibrated.

As a direct read-out device the bridge can be used in a variety of ways. Consider the most general case where all four arms of the bridge are active strain gauges attached to the transducer. It can be shown (Dally and Riley [4]) that the bridge output will be

$$\Delta V = \frac{V}{4} \left[\frac{\Delta R_1}{R_1} - \frac{\Delta R_2}{R_2} + \frac{\Delta R_3}{R_3} - \frac{\Delta R_4}{R_4} \right] \qquad (4.2)$$

Many strain gauge displacement transducers are based on the simple cantilever principle (Fig. 4.12), the required displacement being applied to the free end of the

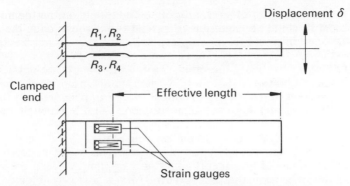

FIG. 4.12   Cantilever displacement transducer

cantilever. With one active gauge, R_1 say, and stable resistors in the other three bridge arms, the transducer would have an output of

$$\Delta V = \frac{V}{4} \cdot \frac{\Delta R_1}{R_1} \, .$$

However, the sensitivity of the bridge can be increased fourfold by using four active gauges located as shown in the illustration, or alternatively increased by a factor of two by using active gauges R_1 and R_3 only.

The versatility of the electrical resistance strain gauge is its chief asset in transducer applications. As an example, quite recently some important recordings were made of the vibration of ducting in a nuclear reactor plant using crude but effective displacement transducers made from ordinary hacksaw blades, strain gauged as in Fig. 4.12, and fixed to a rigid support adjacent to the ducting in question. As in other types of measurement the ability to improvise is desirable and with a little initiative a transducer can usually be devised to provide the required signal. Other factors, such as space limitations, interference effects of the transducer on the deflecting structure, variations in support conditions, and so on, may have to be taken into account.

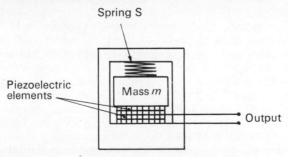

FIG. 4.13   Piezoelectric accelerometer

The range of electrically operated displacement devices is increasing continually and it would be unhelpful to extend the discussion in the present context. However, the section would be incomplete without some mention of a most versatile transducer—the *accelerometer*. There are a number of different designs of accelerometer available, but probably the most common is the piezoelectric compression type. This depends on the principle that when an asymmetrical crystal lattice is compressed an electric charge is produced which is proportional to the applied force. Piezoelectric elements are usually made from barium titanate or lead zirconate.

The accelerometer is shown schematically in Fig. 4.13. The piezoelectric discs are compressed by the mass m which is preloaded by a spring S and the assembly is mounted in a metal container as shown. When the accelerometer is subjected to a vibration, the piezoelectric discs will be subjected to a variable force proportional to the acceleration of the mass m. Due to the piezoelectric effect a variable potential will be developed across the output terminals which will be proportional to the acceleration to which the transducer is being subjected. This variable potential can be monitored on an oscilloscope or on one of the recorders discussed in the previous section. By use of electrical integrating circuits the accelerometer can be used to measure velocity and displacement in addition to the determination of waveform and frequency.

One of the principal advantages of this type of trandsucer is that it can be made so small as to have negligible influence on the vibrating component. A typical size would be 20×16 mm dia. with a total mass of around $0\cdot02$ kg. The frequency range would be typically 2 Hz to 10 KHz.

Measurement of mass, force and torque

The mass of a given body is an invariant, whereas its weight depends on the local gravitational acceleration. Hence the mass of one body is obtained by comparing its weight with that of another body of known mass. This is done by the *mass balance method*, the basic principles of which are universally known. Balances range from the basic equal-arm types to those involving complex lever systems to magnify the out of balance for greater accuracy. For extreme accuracy use can be made of the highly sophisticated balance system developed by the National Physical Laboratory which can make mass measurements of 'kilogrammes with microgramme accuracy'. [1].

Clearly the accuracy of mass balance techniques is directly dependent on the condition of the standard masses used. These should be frequently checked for, (a) excessive corrosion, atmospheric or otherwise and, (b) excessive wear due to contact with other bodies. Many of the more accurate balances now available have the balance weights permanently housed in a sealed unit to minimise corrosion, with external controls to obviate the problem of wear.

Spring balances are commonly used to measure mass, although strictly speaking they are only capable of measuring weight. To be frivolous, one would receive around $0\cdot5$ per cent less value by purchasing say 1 kg of apples at the lowest point on the earth's surface as against the summit of Mount Everest if a spring balance system were used in the purchase.

In the measurement of *force* use is commonly made of linear elastic materials in the construction of the measuring device. The force is determined by applying it to a pre-calibrated elastic element and measuring the resulting displacement. Household scales are an example of this principle, the elastic displacement in this case being converted to a needle rotation for convenience. Probably the most common technical example is the *proving ring* which is basically a high-tensile steel circular ring with loading blocks situated diametrically opposite each other and designed such that tensile or compressive loads can be applied. The change in diameter of the ring gives a measure of the magnitude of the applied load. The diameter change can be measured using a dial gauge, a micrometer head attachment or some form of displacement transducer, e.g. the l.v.d.t. type discussed in the previous section.

Strain gauge type *load cells* also rely on the principle of elastic deformation and, as in the case of displacement measurement, the versatility of the strain gauge allows a broad scope in the type of load cell employed. Table 4.1 indicates the recommended gauge locations for various load cell functions, together with the relevant bridge connections for maximum sensitivity. In each case the gauges are connected such that only the force category required is monitored, e.g. for the measurement of direct loads, equal and opposite bending strains will cancel when connected as shown, and will therefore not affect the bridge balance. Although only simple tensile and bending members are illustrated in the table it should be appreciated that more complex load cells, or *dynamometers* as they are sometimes called, can be designed with the strain gauges situated so as to give various combinations of horizontal and vertical forces, bending moments and torques simultaneously. Octagonal or semicircular ring elements are examples that can be conveniently used in this way.

A convenient device for both applying and measuring loads simultaneously is the hydraulic (or pneumatic) *ram*. In this case the product of effective cross-sectional area and ram pressure gives the applied load. Ram cross-section areas are quoted by manufacturers to within close tolerances and hence the technique can be quite accurate depending on the method of pressure measurement used. The simplicity of the technique, however, can tempt the user to accept the resulting load figures somewhat blindly. Even with careful calibration prior to the application of the test load to account for the friction effects of the hydraulic seals, serious errors can be introduced due to increased friction resulting from unknown side loads. The possible existence of these side loads should be appreciated. In many cases means will be available for eliminating or at least minimising them.

Another hydraulic load measuring device is the *pressure capsule*. The design might involve a flexible diaphragm or a corrugated bellows as part of a closed cell arrangement. Application of load causes a build-up in pressure which can be calibrated to give the desired output. The advantage of this type is that there are no sliding parts and hence friction effects are minimised.

Mention has been made previously of the piezoelectric accelerometer. The same principle is involved in the *piezoelectric force transducer*, the variable potential being developed in this case by the application of the force to be measured rather than the internal mass/acceleration effect of the accelerometer. Piezo-

Table 4.1

Function and gauge layout		Bridge circuits	Comments
1. Axial loads: 2-active gauges 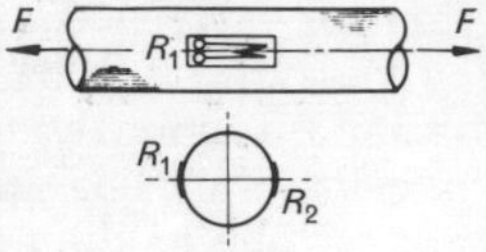	(a) 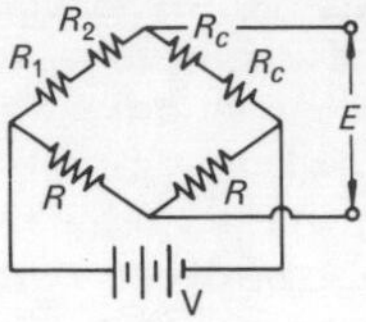		Bending effects eliminated
	(b) 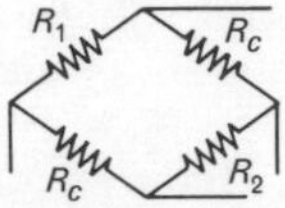		Bending effects again eliminated: sensitivity double that of (a) above
4-active gauges 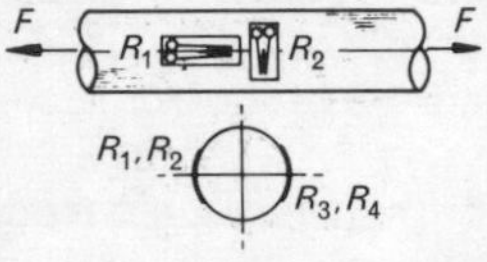	(c) 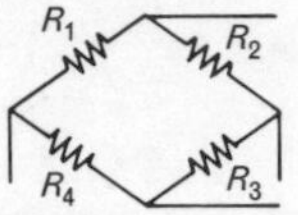		Bending effects again eliminated: sensitivity 2.6 times that of (a) above due to Poisson's ratio effect
2. Bending loads: 2-active gauges 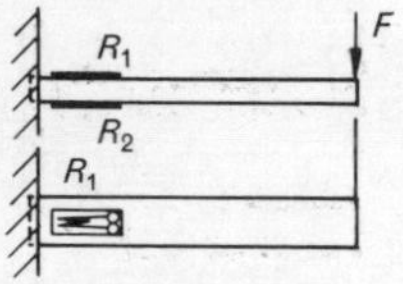	(a) 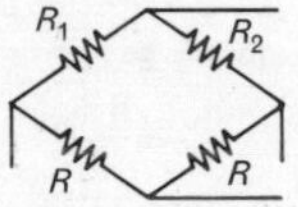		Axial effects eliminated
4-active gauges 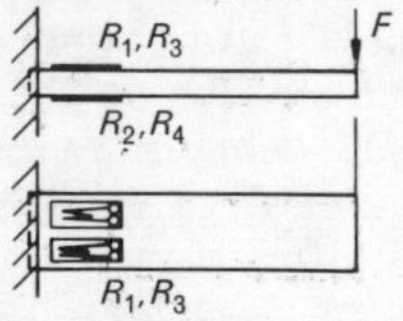	(b) 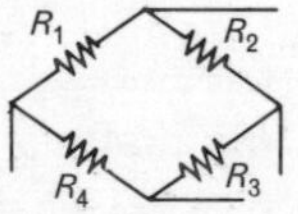		Axial effects again eliminated: sensitivity double that of (a) above
3. Torsion 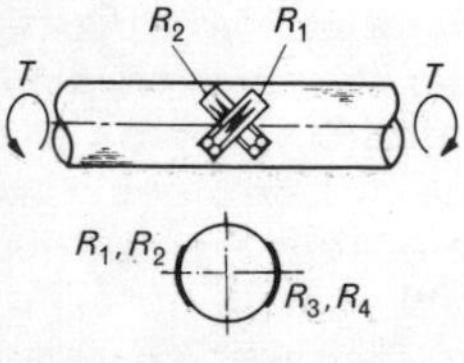			Axial and bending effects eliminated

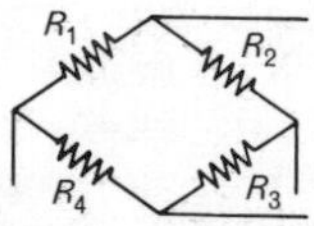

NOTE:

R_1, R_2, R_3, R_4 — Strain gauges
R_c — Temperature compensating strain gauges
R — Stable resistors of equal value

electric force transducers are suitable for dynamic and short-term static applications and a typical range of load capacity would be 15 N up to 1 kN.

Measurement of the power transmitted by a rotating shaft, e.g. from an experimental turbine, or an internal combustion engine, is normally achieved in the laboratory by means of some form of *prony brake*. The power produced is absorbed by a friction device running on a pulley attached to the main drive shaft. The principle is illustrated in Fig. 4.14, which shows the block type prony brake as used for high-speed shafts. The force between the wood blocks and the pulley can be varied as necessary, and by measuring the force F and the lever arm x, the output torque can be determined, from which the power can be deduced. Various designs of prony brake exist, incorporating, for example, belt friction arrangements, different lever systems and in some cases air or water cooling to absorb excessive heat.

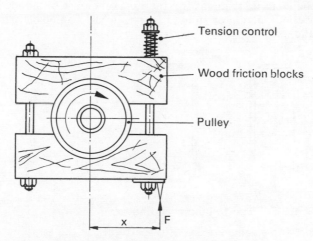

FIG. 4.14   Block type prony brake

The most common form of *torsion dynamometer* is the electric resistance strain gauge type which uses the gauge system indicated in Table 4.1 for torque measurement. Four gauges are attached to the shaft at 45° to the axis, and by connecting them as shown the torque produced can be determined, with any axial and bending strains automatically eliminated. It is necessary, of course, to use a slip-ring arrangement for the bridge connections from the rotating shaft, and this can introduce errors if sufficient precautions are not taken.

No details have been given so far on the problem of the dynamic response of the various load measuring devices discussed. This can be complex and a detailed treatment is not within the scope of the present text. However, the existence of the problem must be acknowledged and reference can be made where necessary to more advanced text-books (e.g. Cook and Rabinowicz [3]).

Measurement of pressure

In a fluid system *pressure* means the force acting upon and normal to unit area of the system. An *absolute pressure* refers to the absolute force acting normally on unit area of the fluid system. It is of interest to note that whilst absolute pressure is always positive for a gas, some rare cases of *negative* absolute pressure have been found for some liquids (Reynolds [8]). *Gauge pressure* represents the difference between the absolute pressure and the local barometric pressure. Very often, in practice, the readings obtained from pressure gauges are relative to the barometric pressure. It is important that the distinction is clearly understood as serious errors in results can sometimes be traced to this point. *'Vacuum'* is the amount by which the barometric pressure exceeds the absolute pressure (i.e. it is then a negative gauge pressure).

Because of its extreme simplicity and reliability, the *liquid-filled manometer* is a popular instrument for the measurement of steady fluid pressures. Many versions of the manometer principle are in use and space permits discussion of only a few of these.

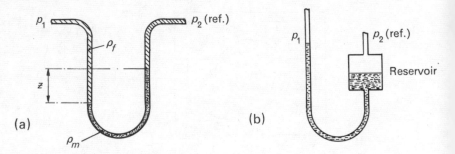

FIG. 4. 15   The U-tube manometer

The *U-tube manometer* (Fig. 4. 15 (a)), is the simplest version and consists of a constant bore glass tube in the form of a long U, partly filled with a suitable liquid such as water, alcohol or mercury. To measure the pressure of a fluid which is less dense and immiscible with the manometer fluid, a connection is made to one limb of the U-tube while a suitable reference pressure is applied to the other limb. The vertical displacement z of the manometer fluid gives an indication of the applied pressure difference Δp. If ρ_m is the density of the manometric fluid, ρ_f that of the fluid whose pressure is to be measured and g is the gravitational acceleration, then

$$\Delta p = (\rho_m - \rho_f)gz \tag{4.3}$$

Inaccuracies may be caused by the effects of surface tension on liquid level. These errors arise if the inside surfaces of the tubes are dirty or their bores are non-uniform. Surface tension effects are greater for tubing of fine bore, and alcohol is often used instead of water as a manometer liquid for gas-pressure measurements, since the surface-tension of alcohol is only about one-third that of water. An important drawback with the use of alcohol in manometers is the gradual change in specific gravity (original value 0·79) because of water absorption. Occasional checks on the specific gravity of alcohol used in manometers lends credibility to the results obtained from them. Note that the surface tension effects of water can be reduced by adding a small proportion of a wetting agent.

Pressures less than about 20 mm water are difficult to measure on a vertical U-tube manometer to an accuracy better than ± 0·5 mm. Sensitivity can be increased by a factor of up to 10 times by inclining the tubes from the vertical. However, angles approaching 5 degrees from the horizontal are not recommended unless the tubes used are very straight. The bore should be about 3 mm for low angles of inclination and fluid of low surface tension used to preserve a satisfactory meniscus shape. Figure 4.15(b) shows an adaption of the simple U-tube which has a very wide internal diameter in one limb. Readings need only be taken from the narrow limb since the fluid level in the reservoir limb remains substantially constant. A correction for differences in surface tension effects may be necessary because of the two different tube diameters.

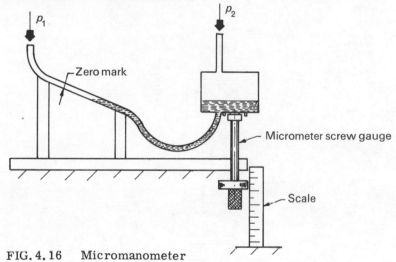

FIG. 4.16   Micromanometer

Compared with the U-tube manometer, the *micromanometer* (Fig. 4.16) can improve measurement accuracy by an order of magnitude but at the cost of some loss of convenience. The reservoir can be adjusted in height by means of a micrometer screw, which changes the level of the liquid in the inclined tube. The inclined tube is marked at one point, which is the *datum*. With an applied pressure difference, $p_1 - p_2$, the reservoir height is adjusted until the level of liquid in the inclined tube is returned to the datum mark. The change in height can be accurately found on the micrometer gauge.

The improved accuracy of this instrument is due to the fact that the meniscus position remains the same for all pressure difference readings. Thus, surface tension effects due to tube non-uniformity or to dirt are reduced. A low-power magnifier is often used to observe the meniscus. The accuracy of the micromanometer can be as good as ±0·002 mm of manometer fluid.

For measuring pressures outside the range of the mercury-filled manometer, the *Bourdon tube,* because of its simplicity and versatility, is the basis of many types of pressure gauge. In its simplest form the Bourdon tube shown in Fig. 4.17 consists of an oval section tube bent into a circular arc. One end is sealed and free to move; the other end is rigidly fixed and is open for the transmission of pressure. By means of a system of linkages a pointer is made to indicate the pressure applied to the tube, as shown in the figure. With some internal pressure the tube section becomes rounder and this causes it to become straighter. It is this motion, amplified mechanically, which is indicated by the gauge needle.

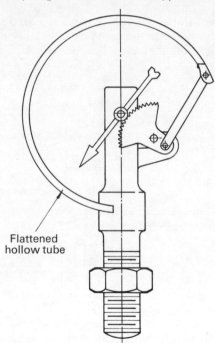

Flattened
hollow tube

FIG. 4.17   Bourdon 'C' type pressure gauge

Pressure gauges are available for the measurement of absolute, 'gauge' or differential pressures over wide ranges. It is essential that periodic checks of the calibration are made to offset the effects of wear on moving parts and ageing of elastic components.

Calibration checks on Bourdon tube gauges are usually performed on a *'dead weight tester'* or hydraulic gauge. Referring to Fig. 4.18, the 'dead-load' W is balanced by a hydraulic pressure acting on a piston of area A. The effects of friction can be almost eliminated by using a low viscosity oil as the hydraulic fluid and manually rotating the piston freely about its axis while a reading is being taken. The device is itself subject to some error due to piston diameter changes caused by alterations in room temperature.

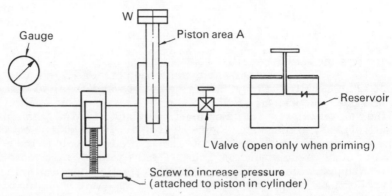

FIG. 4.18

Pressure transducers which convert pressure levels into electrical signals are finding frequent application and can take many forms. Pressure is applied to an elastic element such as a Bourdon tube or a diaphragm or a bellows unit and the resulting movement is measured by some form of displacement transducer as discussed previously (iii) 'Electrical methods', p. 43. In common with other forms of transducer, the pressure transducer lends itself to the measurement of fluctuating pressures. Many different techniques are in use, for example:

(a) resistance changes of strain gauges bonded to a diaphragm;

(b) diaphragm movement as measured by an inductive type displacement transducer or linear variable differential transformer (l.v.d.t.);

(c) capacitance changes produced by the motion of a diaphragm forming one plate of a capacitor. This type is used for measuring rapidly fluctuating pressures which occur in internal combustion engines, for example;

(d) resistance changes by the wiper action of a potentiometer.

Measurements in fluid flow

1. *Velocity*. The velocity V at a point in a constant density fluid flow may be obtained most conveniently from measurements of the *total pressure* p_0

(defined below) and *static pressure p* made at that point. With the fluid density ρ known, the velocity can be determined from Bernoulli's equation, Gibbings [10].

$$p_0 - p = \tfrac{1}{2}\rho V^2 \tag{4.4}$$

The equation can be employed for determining flow velocities in low speed aero-dynamics with only small error up to a Mach number of $0\cdot3$. For Mach numbers M between $0\cdot3$ and unity the compressible flow equation must be employed

$$\frac{p_0}{p} = \left(1 + \frac{\gamma - 1}{2}M^2\right)^{\gamma/(\gamma-1)} \tag{4.5}$$

where γ is the ratio of specific heats of the gas.

On the surface of any solid body immersed in a flowing fluid there is some point, usually the most forward point, at which the fluid is brought to rest. At this point the pressure is the total pressure, sometimes called the *pitot pressure* or the *stagnation pressure*. For low flow velocities ($M \leqslant 1$) this pressure is relatively easy to measure. Figure 4.19(a) shows an instrument called a *pitot tube* which is arranged in a flow so that a small pressure hole at the end of the tube is at or very close to the stagnation point. The instrument will record the total pressure without appreciable error even when not perfectly aligned with the direction of flow. For most purposes it is sufficient to sight the tube within about 5 degrees of the flow direction using some simple device such as cotton thread streaming in the flow. Alternatively, the instrument can be orientated until a maximum reading is obtained (Bryer and Pankhurst [11]).

By enclosing the pitot tube within an open-ended outer tube the pitot reading can be made extremely insensitive to flow direction. This can be very useful in situations where it is not possible to orientate the probe to maximise the reading and the flow direction is uncertain. Figure 4.19(b) shows such a *shrouded total pressure* tube which will remain insensitive to yaw over ±40 degrees.

In supersonic flow ($M > 1$) pitot tubes do not record total pressure correctly. This is because a *detached shock wave* forms upstream of a blunt body in the supersonic stream resulting in some loss of total pressure. The total pressure recorded by the pitot tube is that of the subsonic flow *after* the shock wave. By employing normal shock wave relations (or tables) the free-stream supersonic Mach number can be determined from the pitot tube and static pressure observations.

Although static pressure could be measured by any form of pressure probe if it moved with the fluid, this is seldom possible and a special form of stationary probe is normally used. The accurate measurement of static pressure is inherently difficult because the introduction of the stationary probe into the flow changes the pressure field in its vicinity. The pressure around a probe varies from point to point in a way largely determined by its shape and orientation to the flow direction. The pressure sensing holes on a probe are located where the surface pressure is equal or related in some known way, to the static pressure of the undisturbed flow.

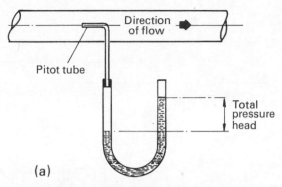

(a)

FIG. 4. 19 (a)   Measurement of total pressure relative to atmosphere

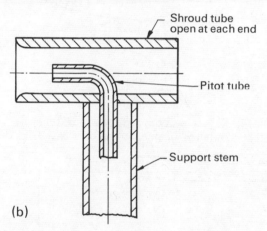

(b)

FIG. 4 19 (b)   Shrouded total pressure tube

A simple type of static pressure probe consists of a body of revolution with its axis aligned to the flow direction, supported by a lateral stem at the down-stream end of the body (Fig. 4. 20 (a)). This particular instrument also includes a forward facing hole for measuring total pressure. Because the flow is brought to rest at the nose the pressure there is greater than that of the undisturbed flow. As the flow accelerates around the shoulder the pressure falls very rapidly below that of the free stream before rising again slowly towards the free stream value. Further along the probe the stem causes a further flow deceleration and the pressure rises above the free stream level.

Figure 4. 20 (b) gives an example of the variation in pressure errors along the probe length due to the separate effects of the nose and the stem. The holes are positioned at the point where the two errors are equal and opposite.

(a)

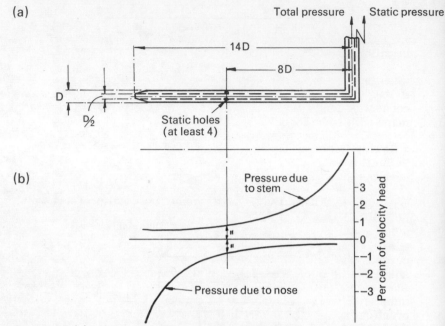

(b)

FIG. 4. 20 (a)   N.P.L. standard pitot static tube
FIG. 4. 20 (b)   Balance of pressures of a N.P.L. pitot static tube

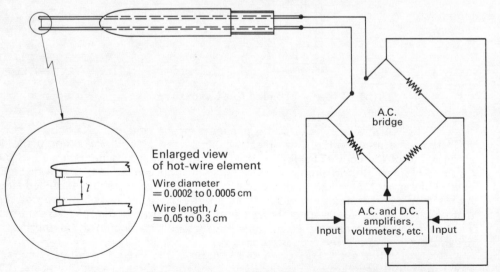

FIG. 4. 21   Hot-wire probe and Wheatstone bridge circuit

This principle is employed in the *National Physical Laboratory (N.P.L.) standard pitot-static tube.* It is often used as a standard of comparison in calibrating other types of pitot-static tubes.

Velocity measurements can only be made with pitot-static instruments when the flow is steady, or very nearly so. The response time of such instruments is long because of the large flow resistance of small diameter pressure holes and connecting leads. In fluctuating flows the measurement of velocity is nearly always made with some form of *hot wire anemometer.*

The temperature attained by an electrically heated wire immersed in a flowing fluid depends upon the velocity of flow past the wire and its orientation to the flow. For a hot wire placed normal to the flow it is possible to calibrate the wire to indicate velocity in terms of wire temperature.

Figure 4.21 shows a simple version of a hot wire anemometer probe which forms, in effect, one arm of a Wheatstone bridge circuit. The wire element is often made of very fine gold-plated or platinum-plated tungsten wire (0·002 to 0·005 mm diameter), soldered or welded to two probe supports (0·5 to 3 mm apart). The bridge supplies current to the wire, heating it to a *constant temperature* between 200° and 300°C. Constant wire temperature, which can be held within very small limits, is made possible by electrical feedback from the amplifier. Thus, when a change in fluid velocity takes place the heating current also varies to maintain the bridge in balance and the wire temperature (or resistance) constant. The practical upper frequency limit of hot-wire anemometers is normally between 5 and 15 KHz depending upon flow velocity, wire size and anemometer characteristics.

Except at very low wind speeds (<5 m/s) the hot wire method of determining mean velocity is less accurate than the pitot-static method mainly on account of its sensitivity to fluid temperature. It also has the serious disadvantages of being fragile and the necessity for individual and fairly frequent calibration. However, for the study of unsteady or turbulent flows the hot wire anemometer is vastly superior to any other system except, perhaps, for the recently developed laser anemometer.

The *laser anemometer* (Fig. 4.22) uses a laser beam (a beam of coherent light) for velocity probing. When the beam passes through the flowing fluid, the light is scattered by suspended particles in the fluid. The scattered light is Doppler shifted and the Doppler frequency is directly proportional to the particle velocity. A photomultiplier is used to pick up and measure the Doppler shifted signal. This measuring technique requires that the fluid flow is translucent and contains particles which scatter light. At the time of writing the performance of one manufacturer's laser anemometer is stated as 3 mm/s to 3 000 m/s velocity range to an accuracy of 1 per cent full-scale deflection of the range in use with an upper frequency limit of 120 kHz and a maximum velocity fluctuation of ±70 per cent from the mean value.

2. *Flow direction.* If pressure holes are disposed symmetrically on either side of a probe, they indicate different pressures when the body is yawed in a fluid flow. This is the principle of the *yawmeter* which is used for measuring flow direction.

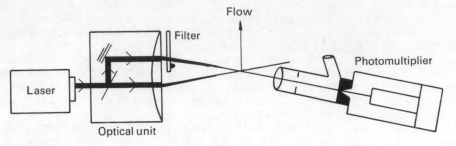

(a) Measurement using reference-beam mode

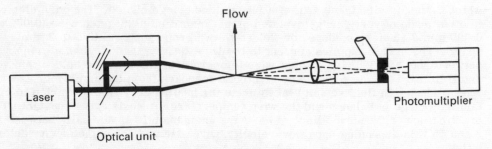

(b) Measurement using the differential-doppler mode on forward scattered light

FIG. 4. 22   Laser anemometer

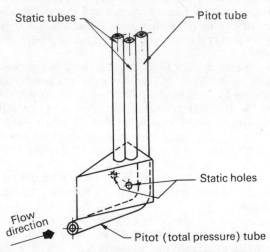

FIG. 4. 23   Wedge Yawmeter and pitot tube

A simple, effective wedge yawmeter, shown in Fig. 4. 23, has a direction sensing hole on either side of a wedge of included angle of about 25 degrees. A pitot tube is included below the wedge making the instrument suitable for combined measurements in a confined space. Some versions also include a small thermocouple.

In use, the orientation of the wedge is adjusted until the readings from the two faces are equal and the yaw angle is read off an accurate protractor head attached to the stem. The readings from the wedge faces are equal to the sum of the static pressure and a constant fraction of the dynamic pressure, $\frac{1}{2}\rho V^2$. In order to find the static pressure in a flow using a wedge yawmeter it is necessary for a calibration to be made of the indicated dynamic pressure of the instrument with that of another instrument (e.g. the N.P.L. standard pitot-static tube).

3. *Mass flow*. Although wedge yawmeters are suitable for detailed flow exploration and can, in principle, be used for measuring overall mass flow through a duct by traversing them across the duct, this would be tedious and rather impractical. The simplest method of measuring the flow through a duct is to pass the whole flow through some sort of constriction and measure the pressure drop across it. Figure 4. 24 shows two commonly used flowmeters which are designed on this principle.

The *orifice plate flowmeter*. Fig. 4. 24 (a), is relatively crude but inexpensive to install. It is mostly used when the irrecoverable pressure drop caused by the plate is unimportant. The *venturi tube* (Fig. 4. 24 (b)) causes little

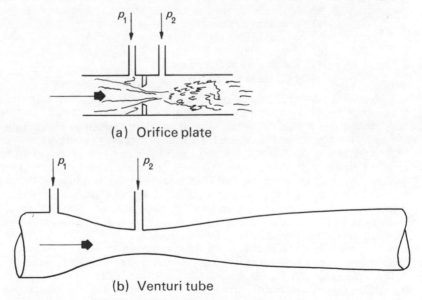

(a) Orifice plate

(b) Venturi tube

FIG. 4. 24   Two common flowmeters

loss in total pressure because of the smooth deceleration of the flow after the con-
traction; however, it is more expensive.

When these flowmeters are made to standardised proportions, the
pressure difference measured, $(p_1 - p_2)$, can be very large and may be measured
to an accuracy better than 99 per cent without using a manometer of high sensitivity.

The average velocity can be determined from a measurement of the
centreline velocity and a knowledge of the velocity profile across the duct. This
is not a particularly good method and is not recommended, as it is sensitive to
asymmetry and flow perturbations. However, a modification of this simple idea
which is recommended (as it is cheap and reasonably accurate) is the *three-quarter
radius flowmeter*. This consists simply of four pitot tubes at intervals of 90 degrees
around a circle co-axial with the duct axis and of radius three quarters of that of
the duct radius itself. Four equispaced wall static pressure orifices are also used.
The three-quarter duct radius was chosen since this is the position at which the
velocity is a constant fraction of the average velocity over a wide range of Reynolds
number. One practical difficulty is that the flowmeter requires an upstream
'settling length' of at least 10 duct diameters.

Measurement of temperature

The temperature of an object is a measure of the thermal motion of
the molecules or atoms comprising the object. All temperature-measuring devices
record some physical manifestation of this thermal motion, e.g. the change in:
(a)   physical state;
(b)   chemical state;
(c)   dimensions;
(d)   electrical properties;
(e)   radiation properties.

The best known temperature measuring device is the *liquid-in-glass*
thermometer. This relies on the relative expansion between the two substances, e.g
the coefficient of cubical expansion of mercury is about eight times that of glass.

It is not often realised that a mercury-in-glass thermometer, which
is in apparently good condition, may have a large error due to aging. When glass
is heated to a high temperature it does not immediately contract to its original
volume on cooling. Thus, at low temperatures the reading on the thermometer will
be too low; it normally returns to its original size after a period of time. The duration may be of the order of minutes and depends upon the nature of the glass from
which the bulb is made. Calibration against a standard (e.g. the melting point of
ice) is really required at frequent intervals.

For measuring high temperatures (mercury boils at 357 °C at atmospheric pressure), the *top* end of the thermometer capillary tube is enlarged into a
bulb. This bulb has a capacity of about 20 times that of the capillary tube itself.
By filling this volume with nitrogen or carbon dioxide at 20 atmospheres the range.
of the thermometer can be extended to above 500 °C.

Most liquid-in-glass thermometers are calibrated in a bath so that the liquid in the capillary tube is totally immersed. For the user it is very convenient to have a length of the liquid column exposed to view so that readings can be readily taken. However, this practice may expose the emergent liquid to an appreciable temperature difference and give a reading too low or too high depending on whether the ambient temperature is lower or higher than the substance being measured. A simple correction can be applied (Benedict [2]) with the equation

$$\Delta T = KN(T_1 - T_2) \tag{4.6}$$

where ΔT is the *stem correction* in degrees, K is the differential coefficient of expansion, N is length of exposed thermometric fluid in degrees, T_1 is the bulb temperature and T_2 is the average temperature of emergent liquid.

It is easy to acquire a too-casual attitude to readings taken from liquid-in-glass thermometers, perhaps because they are in everyday use. Serious errors may, however, be caused by the stem effects as the following example will show.

Example. A mercury-in-glass thermometer indicated 400 °C when the column of mercury was immersed in a liquid up to the 90 degree mark. If the mean temperature of the mercury not immersed was 100 °C what was the true temperature of the liquid? The differential expansion coefficient between the mercury and glass is $1.6 \times 10^{-4}/°C$.

$$\Delta T = 1.6 . \, 10^{-4} \, (400 - 90)(400 - 100) = 15 °C$$

This is a first approximation as the indicated temperature was used for T_1. With $T_1 = 415 °C$ a second approximation gives $\Delta T = 1.5 . \, 10^{-4} \, (400 - 90)(415 - 100) = 15.6 °C$.

Thus, in order to correct such large errors it is necessary to determine the temperature of the vapour (or gas) using a secondary thermometer.

The use of liquid-in-glass thermometers is often limited by (a) the fragility of the glass, and (b) the difficulty of reading the scale because of inaccessibility. The *mercury-in-steel thermometer* (Fig. 4.25) overcomes these limitations. Readings are obtained by pressure transmitted through a capillary tube (steel) to a Bourdon pressure gauge. As the capillary tube can be long, (e.g. 40 m or so) the thermometer can be read some distance away from the bulb. An allowance should be made for the difference in height between bulb and dial. This type can work up to about 650 °C with an accuracy of about 2 per cent.

The difference in the expansion coefficients of two metals can be used to indicate changes in temperature. The two metals are welded or brazed together along the surfaces. Typical metals used are invar and brass. When heated, the brass expands faster than the invar, causing the strip to deflect. Long *bimetal strips* wound into the form of a helix can be used as the temperature sensing element of thermometric devices.

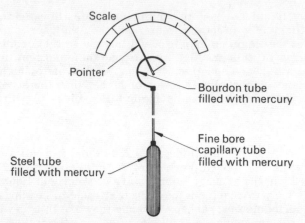

FIG. 4.25   Mercury-in-steel thermometer

Resistance thermometers comprise a 'sensor' whose electrical resistance varies with temperature, a means of measuring resistance, and a relation between resistance and temperature. Typically, a resistance thermometer sensor consists of a fine platinum wire wound on to a mica frame and enclosed in a protective tube. The leads are taken to some form of bridge circuit by which changes in resistance with temperature are measured with high accuracy. Features that characterise resistance thermometers are stability of sensors, sensitivity of measurements and simplicity of the circuitry.

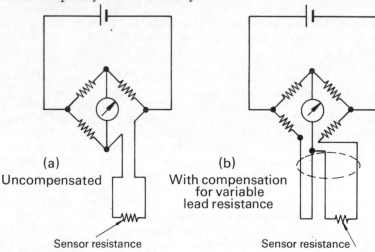

N.B. Thermo-electric effects can be eliminated using A.C. instead of D.C. excitation

FIG. 4.26   Resistance thermometers as part of Wheatstone bridge circuit

Figure 4.26 shows resistance thermometer sensors as part of Wheatstone bridge circuits suitable for normal engineering laboratory use. For high precision the conventional Wheatstone bridge is not very satisfactory since:

(a) slide-wire contact resistance is a variable dependent upon condition of slide;

(b) temperature gradients along extension wires to sensor cause variable resistance of wires;

and (c) supply current itself causes variable I^2R heating due to changes in sensor resistance.

Benedict [2] gives details of various refinements to circuits to alleviate these difficulties.

In 1821 Thomas Seebeck discovered that when the junctions of two dissimilar metals comprising a closed circuit were exposed to a temperature difference, a net e.m.f. was produced which induced a continuous electric current around the circuit. The phenomenon is called the *Seebeck effect* and involves the conversion of thermal energy into electrical energy. The subject of *thermoelectricity* which deals with the theory of the Seebeck effect and related phenomena is dealt with in numerous thermodynamic texts, e.g. Soo [9], Benedict [2].

Thermocouples, which are extremely useful thermometers in many research and engineering applications, employ the Seebeck effect under zero current conditions to indicate the temperature of one junction relative to that of the other. Figure 4.27 shows a simple thermocouple electric circuit in which the cold junction is maintained at a constant reference temperature of 0 °C by a bath of melting ice. The e.m.f., V volts, is usually measured on a potentiometer. The thermocouple can be calibrated by measuring the e.m.f. at various known temperatures with the reference junction maintained at 0 °C. For most practical purposes

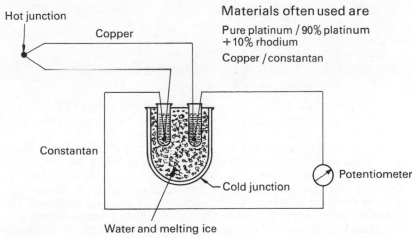

FIG. 4.27   Thermocouple

the results of such measurements can, for most thermocouples, be represented with sufficient accuracy by the quadratic equation

$$V = A + BT + CT^2 \qquad (4.7)$$

where A, B and C are constants relating to a particular thermocouple and T is the temperature.

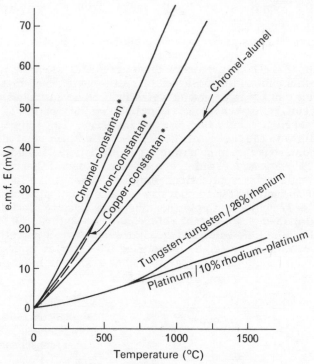

* Can be used down to −180°C

FIG. 4. 28   Thermocouple output voltage against temperature for various material combinations (Reference Junction at 0 °C)

There are a large number of possible material combinations in thermoelectric thermometry but only a few of these are actually used in practice. These are chosen on the basis of thermoelectric potential, Seebeck coefficients, stability and reproducibility. Figure 4. 28 shows the Seebeck voltage plotted against temperature for a selection of typical material combinations. The end of each curve approximately indicates the upper limit of temperature for a particular combination. Some of these are marked with an asterisk to indicated they can be used also at low temperature (−180 °C).

Radiation detectors are used for measuring both radiant energy and temperature. They fall into two broad groups depending upon whether the radiation is measured over a wide band or a narrow band of wavelengths. In the case of the *total-radiation thermometer*, radiation emitted from a surface is focused onto a temperature sensitive detector such as a thermocouple and the resulting electrical signal used to indicate temperatures up to about 3 000°C

Selective radiation thermometers, usually known as *optical pyrometers*, work as brightness comparators. One common type of optical pyrometer (Fig. 4. 29) consists essentially of a telescope containing a narrow band wave filter (usually red glass) and a tungsten filament electric lamp. When the current supplied to the lamp is gradually increased, the filament viewed against the brightness of the radiating body tends to disappear. A further increase of current will cause the filament to glow more brightly than the background. If the relationship between heating current and filament temperature is known the temperature of the radiating surface can be determined. Commercial optical pyrometers often have the ammeter graduated to give a direct temperature reading.

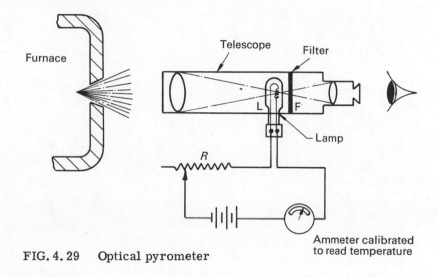

FIG. 4. 29   Optical pyrometer

The disappearing filament optical pyrometer can be used to measure the temperature of radiating bodies in excess of 750 °C up to the limit of the lamp filament at 1 350 °C. When temperatures in excess of 1 350 °C are to be observed, glass filters are interposed between the lamp and the radiating source to absorb some of the radiation. Temperatures in excess of 4 000 °C can be measured in this way.

The main source of error of all radiation thermometers is caused by uncertainty in the emissivity of the source.

Table 4.2 Summary of commonly used laboratory equipment

Parameter to be measured	Measuring device	Typical range	Approximate resolution	Remarks
LENGTH	Machinist's calipers with steel rule	0 to 50 mm up to 0 to 1·0 m	±0·2 mm	Used mainly for checking nominal dimensions.
	Vernier calipers or height gauge	0 to 0·1 m up to 0 to 0·5 m	±0·02 mm	
	Micrometer	0 to 25 mm or 25 to 50 mm etc.	±0·005 mm	Range can be extended considerably to, say, 1 m ±12·5 mm by use of a standard 25 mm micrometer head and a 'bow gauge' structure.
	Slip gauges	e.g. 0·5 to 100·0 mm	±0·001 mm	Used mainly for the calibration of lower precision instruments.
DISPLACEMENT	Dial gauge	e.g. 0 to 5 mm 0 to 50 mm	±0·01 mm	Extensively used, particularly in workshop practice.
	Linear variable differential transformer	e.g. 0 to 0·25 mm up to 0 to 0·3 m	±0·1%	Resolution theoretically infinite but in practice dependent on associated instrumentation.
ACCELERATION	Accelerometer	e.g. 0-7 000 g	0·05 g	With suitable instrumentation can be used for recording periodic displacements
STRAIN	Mechanical extensometers	e.g. Huggenberger type 0 to 0·4%	± 10 ×10⁻⁶	Range can be increased by resetting.
	Electrical resistance strain gauges	±1·5% up to ±20% strain	±1 × 10⁻⁶	Generally useful in transducer applications for measurement of displacement, load, torque pressure, etc. Not re-usable and therefore costly.
TIME AND FREQUENCY	Stop clocks	e.g. 0-60 min	±0·2 s	Not suitable for short periods of time, say < 10 s. Stop watches available with better resolution.
	Tachometer (machanical or electrical)	e.g. 0-50 000 rpm	—	

	Instrument/Method	Range	Accuracy	Comments
	Stroboscope	e.g. 0-100 Hz 0-300 Hz	—	Especially useful for shaft speed or frequency measurement, where it is undesirable to have direct contact with the moving part.
	Cathode ray oscilloscope (c.r.o.) with transducer	e.g. 0.5 µs to 1 s d.c. to 3 MHz	±5% —	By far the most widely used item of equipment for examination of transient and repetitive signals.
	Electronic frequency counters with transducer	e.g. 1 µs to 10^4 s 10 Hz to 1.2 Hz	e.g. ±1 µs —	Very versatile instrument for use with any form of wave or pulse pickup. Can be used for time or frequency measurement.
	Trace recorders with transducer	0 to 2 Hz up to 0 to 5 kHz	±1%	Pen recorders are used for low frequency response; ultra-violet recorders, with suitable galvanometers for high frequency response. Adjustable chart speed.
MASS	Balances	e.g. 0 to 20 g → 0 to 200 g → 0 to 5 kg →	±0.001 mg ±1 mg ±4 mg	For extreme accuracy 'ultra-micro-balances' are available with a resolution of ±0.1 µg.
FORCE	Spring balance	e.g. 0-20 N up to 0-5k N	±1.0%	Resolution often poorer in the simpler spring balance arrangements.
	Proving rings	e.g. 0 to 0.2 kN up to 0 to 70 kN	±0.1%	Accuracy dependent on technique used for measuring ring deflection.
	Strain gauge load	e.g. 0 to 40 N up to 0 to 1 MN	±0.1%	If not available commercially, strain gauge load cells can usually be devised to provide the required force measurement. (See Table 4.1.)
	Hydraulic or pneumatic ram	Any range up to about 2 MN	See comments	Accuracy very much dependent on friction effects, but also on pressure measuring system.
	Piezoelectric force transducer	e.g. 0 to ±8 kN up to 0 to ±5 MN	±20 mN	These transducers noted for their high resolution.
TORQUE	Prony brakes	General principle suitable for most laboratory power units	see comments	Used for power measurement in rotating shafts. Accuracy dependent on force measuring technique, etc.
	Torsion dynamometer	0 to 10 mNm up to 0 to 5 kNm	±0.1%	This type suitable for either 'static' or 'rotating' torque measurement

Table 4.2 continued

Parameter to be measured	Measuring device	Typical range	Approximate resolution	Remarks
PRESSURE	U-tube manometer	20 mm to 5 m of manometer fluid (vertical U-tube)	±0·5 mm ±0·05 mm at 5 deg to horizontal	Range can be increased but is not usually practicable or convenient. Lower limit of range can be reduced to about 2 mm by tilting tubes to 5 deg. from horizontal.
	Micromanometer	0 to 0·2 m of manometer fluid	±0.002 mm	A high precision instrument.
	Bourdon tube	0 to 0·1 MN/m^2 up to 0 to 500 MN/m^2	±1%	Combined negative and positive ranges sometimes used on low pressure gauges.
	Dead-weight tester	100 N/m^2 to 100 MN/m^2	±0·01 to ±0·05%	Precision depends upon quality of device. Used for calibrating other gauges.
	Pressure transducers	0·002 mm of water to 500 MN/m^2	±0·1%	Sensitivity to low pressure differences limited by friction or by thermal expansion.
FLOW VELOCITY	Pitot-static tube	6 to 60 m/s usually	±1% of dynamic pressure	Quoted accuracy applies to International Standard designs. Possible to extend down to 1 or 2 m/s. Can be used up to Mach 5.
	Hot-wire anemometer	0·1 to 250 m/s in filtered air at temperatures up to 150 °C	See comments	Less accurate than pitot-static method except at very low velocities. Has very fast response—is very good for unsteady flow and turbulence measurements. Fragile.
FLOW DIRECTION	Wedge yawmeter	—	±0·2 deg or less	Also measures velocity. Works well in transverse gradients of total pressure. Small size.
MASS OR VOLUME FLOW RATE	Orifice plate	See remarks	±1%	Simple, cheap but has large losses in pressure. Size of plate or tube chosen to give a differential pressure of 0·5 to 5 m of fluid on a U-tube manometer.

	Instrument	Range	Accuracy	Remarks
	Venturi-meter	See remarks	±1%	Fairly expensive but has low pressure losses.
	¾ Radius flow meter	Minimum velocity about 0·5 m/s for water and 12 m/s	±½%	Has low pressure losses. Can be damaged or blocked by particles in fluid flow.
TEMPERATURE	Mercury in glass thermometer Alcohol " " Pentane " "	−35 to 510 °C −80 to 100 °C −200 to 30 °C	±½°C	High temperature can cause aging in glass. Zero calibration needed (in melting ice).
	Mercury in steel thermometer with Bourdon gauge	− 35 to 650 °C	±1%	Useful for remote readings. Correction for differences in level required.
	Bimetal thermometer	−180 to 450 °C	±1%	Slow response.
	Platinum resistance thermometer	−240 to 1 060 °C	±0·2 °C ($0 \leqslant T \leqslant 100$ °C) ±0·35 °C (at 200 °C) ±0·55 °C (at 300 °C) ±0·8 °C (at 400 °C)	Very high precision—often used for interpolating between primary standard temperature points. See BS 1904 for temperature-resistance relationship and tolerances.
	Thermocouples Chromel & constantan Copper & constantan Iron & constantan Chromel & alumel Tungsten & Rhenium Platinum & rhodium	−180 to 1000 °C −180 to 400 °C −180 to 850 °C 0 to 1100 °C 0 to 2800 °C 0 to 1450	±5°C ±10°C ±10°C ±3 °C	Simple and cheap Use in non-oxidising atmosphere. Not subject to corrosion—very reliable—expensive.
	Optical pyrometers (disappearing filament type)	750 to 4 000 + °C	±4 °C (at 1 000 °C) ±6 °C (at 2 000 °C) ±40 °C (at 4 000 °C)	When calibrated against a standard. Glass absorption filter used above about 1350 °C.
VOLTAGE AND CURRENT	Moving coil and moving iron instruments etc.	See comments	±0·5% for precision grade ±1% to ±3% for industrial grade	Lower limit of effective range varies according to type of movement (typically this is 10% to 30% of full scale). See BS 89 for further information.

General note. Any one instrument may cover only part of the ranges shown in the table. Ranges and resolution quoted are for guidance and may differ between manufacturers of the same type of instrument.

Summary of commonly used laboratory equipment

In Table 4.2 the various items of equipment mentioned in this chapter have been tabulated, together with their range and resolution. For further details on any one item reference should be made to the foregoing text or to the references listed.

References

1. Anonymous. Balances, Weights and Precise Laboratory Weighing, *Nat. Phys. Lab. Notes on Applied Science,* No. 7, 1962.
2. R. P. BENEDICT, *Fundamentals of Temperature, Pressure and Flow Measurement.* Wiley, 1969.
3. N. H. COOK and E. RABINOWICZ, *Physical Measurement and Analysis.* Addison-Wesley 1963.
4. J. W. DALLY and W. F. RILEY, *Experimental Stress Analysis.* McGraw-Hill, 1965.
5. J. C. EVANS and C. O. TAYLORSON, Measurement of Angle in Engineering, *Nat. Phys. Lab. Notes on Applied Science,* No. 26, H.M.S.O., 1964.
6. K. J. HUME, *Engineering Metrology.* Macdonald and Company, 1950.
7. H. K. P. NEUBERT, *Instrument Transducers.* Oxford, 1963.
8. O. REYNOLDS, 'On the Internal Cohesion of Liquids and the Suspension of a Column of Mercury to a Height more than Double that of the Barometer', *Mem. Proc. Manch. lit. phil. Soc.,* 3rd Series 1882, Vol. 7, p. 1.
9. S. L. SOO, *Direct Energy Conversion.* Prentice-Hall, 1968.
10. J. C. GIBBINGS, *Thermomechanics.* Pergamon Press, 1970.
11. D. W. BRYER and R. C. PANKHURST, *Pressure-Probe Methods for Determining Wind Speed and Flow Direction.* H.M.S.O., London, 1971.

Further reading

P. BRADSHAW, *Experimental Fluid Mechanics.* 2nd Edition, Pergamon, 1970.
B. J. BRINKWORTH, *An Introduction to Experimentation.* English Universities Press, 1968.
J. P. HOLMAN, *Experimental Methods for Engineers.* McGraw-Hill, 1966.
E. B. JONES, *Instrument Technology,* Butterworths, 2nd Edition, 1965.
E. OWER and R. C. PANKHURST, *Measurement of Air Flow.* Pergamon, 4th Edition, 1966.

5   *Errors in Experimentation*

Every measurement involves an error. The nature of errors may vary and so may their magnitudes but total elimination of errors from experimentation and testing remains beyond human power.

Irrespective of its formulation or nature, error is the deviation of a measured value from the corresponding true value. Errors comprise a number of constituent components. Some of them can be eliminated completely by special care and attention, and some cannot.*

Because errors cannot be avoided in general one must learn how to live with them. One must also be able to assess their magnitude and, moreover, to be able to control their magnitude according to justified needs.

Errors are frequently random or include a random component. Therefore statistical methods and the probability theory play an important part in the analysis of errors. To be able to predict the effects of errors upon measurements and/or processes is often absolutely vital. Engineers, it is sometimes said, are on average less error conscious than physicists. If this is true, there is obviously a cause for concern.

The relevant literature on experimental errors is extensive and it would be difficult to list it in detail in this chapter. It may be, however, worth noting that the following textbooks: Beckwith and Buck [1], Brinkworth [4], Cook and Rabinowicz [5], Richards [8], could be valuable as supporting reading. Further details on the analytical treatment of data can be found in: Pugh and Winslow [7], Beers [2], Moroney [6], and Braddick [3].

---

* There is one exception to this, namely counting, which is often classified as the simplest form of measurement. Counting of numbers, especially of small numbers, can be executed without any error at all.

5.1   Some useful definitions and concepts of error quantities

Absolute error

The *absolute error* E is equal to the actual difference between the true value X_T and the corresponding measurement (observation) X_M, or calculation, if such a calculation involves measured quantities or/and approximations.

Absolute Error $E = X_T - X_M$

Relative error

A nondimensional form is obtained by dividing the absolute error by its measured value. Such a form is known as the *relative error*.

$$\text{Relative Error } e = \frac{E}{X_M} = \frac{X_T - X_M}{X_M} = \frac{X_T}{X_M} - 1$$

Tolerance

The *error tolerance* is the region within which the true value can be found. If δ_{max} and δ_{min} denote the maximum and minimum error limits then

$$X_T = X_M \begin{array}{l} + \delta_{max} \\ - \delta_{min} \end{array}$$

or

$$X_M - \delta_{min} \leqslant X_T \leqslant X_M + \delta_{max}$$

In the case of symmetrical tolerance, $\delta_{max} = \delta_{min} = \delta$, one writes

$$X_T = X_M \pm \delta$$

For example, $51 \cdot 3^{+0 \cdot 2}_{-0 \cdot 4}$ means that the true value $51 \cdot 3$ lies between $51 \cdot 5$ and $50 \cdot 9$. An experimental value measured to an accuracy of 5 per cent may be denoted as

$$X_T = X_M (1 \pm 0 \cdot 05)$$

Decimal accuracy

In the experimental world the quantity (for example) $51 \cdot 3$ mm never signifies the ideal value $51 \cdot 300000\ldots$ mm to an infinite number of significant places. It means, according to the convention adopted, that the result is closer to the precise value $51 \cdot 30000\ldots$ mm than to the limiting precise values $51 \cdot 20000\ldots$ mm or $51 \cdot 4000$ mm. Similarly, the measurement $51 \cdot 30$ mm should be understood to be closer to the precise value $51 \cdot 3000\ldots$ mm than to the precise values $51 \cdot 29000\ldots$ mm and $51 \cdot 31000\ldots$ mm. In the former case the measurement was taken with an accuracy of $1/10$ mm (ordinary vernier scale) and was found to lie between $51 \cdot 25$ and $51 \cdot 35$ mm, whereas in the latter case the same measurement was accurate up to $1/100$ mm (micrometer) and the reading lay between $51 \cdot 295$ and $51 \cdot 305$ mm.

Rounding off

An observer often wishes to round off his experimental results or to approximate his calculations on account of the uncertainty of the last figures of the result. One tends to keep two uncertain figures in any intermediate reading, observation or result, but to restrict the final result to only one doubtful figure. As an example imagine a U-tube mercury manometer equipped with a vernier scale capable of providing readings with an accuracy of $\frac{1}{10}$ mm. A routine reading was taken and recorded as 1 235·4 mm Hg. However, it was noticed that some low frequency pressure fluctuations of the order of ±1 mm Hg were present, making the last two figures of the above result uncertain. The question now arises how should one round off the reading in order to provide reliable information?

There is a number of rounding off rules whose primary purpose is to reduce the cumulative error. The one that is recommended below is probably most common.

Increase the last retained digit by one if the adjacent digit to be dropped is:

1. $>$ 5,
2. $=$ 5 and is followed by digits $>$ 0,
3. $=$ 5 and is not followed by significant digits, or followed by zeros only, and preceding digit when increased by one becomes even.

Examples:

13·36|712 → 13·37; 13·36|513 → 13·37; 13·37|5 → 13·38;
13·37|500 → 13·38; 13·36| 50 → 13·36; 13·36|412 → 13·36.

Accuracy and precision

In everyday language the two words 'accuracy' and 'precision' appear to be almost synonymous. Different meanings, however, are usually given to these words in the error analysis.

The word *accuracy* is reserved to refer to systematic errors, whereas the word *precision* is related to all incidental, i.e. random, errors. A miscalibrated instrument (e.g. with a displaced scale) can be read very precisely by a meticulous observer but the results obtained could be inaccurate.*

---

* A house owner, who happened to be an engineer, decided to check functioning of his electricity meter installed by an Electricity Board. By running just one 1 kW electrical appliance (he made sure that all other appliances had been turned off) for exactly one hour he found, to his discomfort of course, that the meter recorded 1·37 kWh instead of exactly 1·00 kWh. The engineer repeated the experiment several times and very much the same result was obtained. His prompt complaint to the Electricity Board stated that the meter installed was *precise,* all right, but very *inaccurate* to his disadvantage. The Board sent a man to replace the meter and our engineer again applied his own test five times to find the readings were this time 1·2, 0·90, 1·0, 1·10 and 0·80 kWh. The precision of the new measurement was poor although its average reading was 1·00 kWh. Incidentally, our engineer did not complain again—he gave up!

In other words accuracy means correctness of measurements and precision means consistency of measurements.

5.2   Classification of errors and their nature

A remark was made in the previous paragraph that errors are generally divided into two classes: (a) *Systematic* (accountable, or fixed) errors and (b) *Random* (unaccountable, or chance) errors.

Systematic errors

The errors belonging to this class are:
1. *Method errors:* These errors arise when a wrong or insufficient experimental method has been chosen. Measuring of one quantity in mistake for another may occur or some unrecognised effects may influence the quantity measured so that the resultant values become erroneous. Unjustified extrapolation of experimental data may also lead to method errors.
2. *Instrument errors:* Errors of this type can be caused by a faulty instrument, the misoperation of an instrument, or by using an instrument in the environment for which it was not designed. The instrument errors are frequently biased in one direction, although in some situations hysteresis effects can occur (e.g. a worn out micrometer, or traverse mechanism).
3. *Calibration errors:* Most instruments will not yield correct results unless they are calibrated before use against a known quantity. This may involve a simple zero setting or determination of a whole calibration curve (or scale). In either case errors can creep into the calibration procedure.
4. *Human errors:* Human errors depend on the personal characteristics of the observer. A human may respond to a signal too early or too late; he may either overestimate or underestimate the reading. Such errors are usually fairly consistent as they are committed perpetually by the same observer at a single session. Occasional errors committed spasmodically, owing to relaxation of vigilance for example, do not apply here and are classed as mistakes.
5. *Arithmetic errors:* Arithmetical calculations involved in experimentation are nowadays being increasingly taken over by various automatic computing devices (computers, automatic desk calculators, slide rules, etc.). Aberrations of such devices, however infrequent, cannot be ruled out completely. Additionally, there may be faults in the actual calculation procedures (programs). Incorrect rounding off can also contribute to arithmetical errors.
6. *Dynamic response errors:* It is perhaps slightly out of place to devote a separate section to the dynamic response errors. However, their significant participation in modern experimentation, especially in connection with measuring time-dependent variables, warrants a separate emphasis. Unlike static response errors (static nonuniformity of action, hysteresis), dynamic response errors arise when an instrument recording a fast changing signal fails to respond linearly to the signal variation (e.g. a pitot-static tube in fluctuating fluid flows, or electrical instruments applied to nonsinusoidal electrical currents, etc.)

Systematic errors are generally avoidable in so far as they can be detected, prevented or allowed for. Before planning an experiment one must consider all the potential sources of systematic errors and be aware of their inherently cumulative nature.

Random errors

Random errors are different from those already discussed. They are not systematic but subject to irregular, chance or random causes. Random errors are the most difficult to account for although their contributions to the overall error level may be considerable and often dominant.

The most common types of the unaccountable random errors are:

1. *Mistakes or errors of judgement:* If a steady atmospheric pressure is displayed on a standard mercury manometer and ten different observers take one careful reading each, then even after elimination of all systematic errors (or after reducing them to a minimum), not all the relevant readings will be the same. People tend to judge differently. Moreover, the same person can judge differently the same quantity on two different occasions. Further, there are genuine mistakes, however sporadic they may be. Uncontrollable human factors such as sudden distractions, tiredness, misunderstandings, etc., affect the correctness of readings and record taking. Calculations are also subject to mistakes.

2. *Variation of conditions:* This is the second important cause of random error. A sudden and unexpected flow disturbance may temporarily alter a thermometer reading in a pipe flow. If it happens that a reading is taken at this particular time the results will be subject to a chance error. Similarly a noise-measuring microphone can suddenly and unexpectedly pick up extra sound from a passing aircraft. Generally such error sources are of limited duration or are due to specific environments.

3. *Specification errors:* An intermediate inspection of a crankshaft main bearing journal in an automobile plant, for example, can be easily subject to an error if the specification for measuring its diameter is not strict enough in limiting the measurement to one position. The journal, even after grinding, may be inconspicuously conical, oval, or contain flats, and even though the measuring system is correct the results may prove scattered. From the production standpoint, however, this scatter may be quite acceptable as long as it is within the prescribed tolerance.

Having described some of the possible sources of random errors the remaining question is: how does one account for random errors? The general tendency is to employ statistical methods. One looks at the distribution of random errors and the probability of their occurrence in order to be able to control the precision of experimentation and to be able to increase it, if need be. In Appendix 2 some elementary rules will be presented concerning the statistical processing of errors and the relevant interpretation. No attempt will be made here to provide an exhaustive and extensive treatise on the applied statistics and the literature should be consulted for further details. Beginners may find Moroney [6] useful as an introductory reading.

General remarks

Yet another group of errors is sometimes quoted in the literature: illegitimate errors. These are simply blunders and chaotic errors and it is best left to the reader to decide how to deal with them.

Two important categories of experimental observations should be distinguished:

1. *Multisample measurements:* mean repetitive measurements of a certain quantity under varying test conditions (different observers and/or different instruments).

2. *Singlesample measurements:* imply one reading or succession of readings executed under identical conditions but at different times.

They refer to the problem of clearing doubts about the experimental results. A repeatability test is one way of gaining confidence, but a far more reliable way is to use an entirely different method to obtain the same results or to support a conclusion.

Having defined and classified the experimental errors, as well as having commented on their physical implications, the next step is to employ numerical procedures to quantify them. As some of the procedures, although by no means difficult, may be beyond the grasp of an inexperienced reader, it has been decided to present the respective material in Appendix 2. The introductory reading of this textbook should thus remain uninterrupted by analytical discourses. On the other hand a reader who wishes to use the present textbook as an experimental manual has the benefit of Appendix 2, where ready-to-use technical information and practical recommendations concerning the rudiments of error analysis are provided. The appendix deals with the following items:

Forms of presentation of error affected data;
Evaluation of such basic parameters as: the arithmetical mean, the standard deviation, etc.;
The Gaussian law of errors,
Suggestions how to reject dubious data;
Testing of experimental data for deviations from the normal (Gaussian) distribution;
Propagation of errors;
Curve fitting.

The supporting literature provided in Chapter 5 applies directly to Appendix 2.

The material presented in Chapter 5 and Appendix 2 is meant to be for general guidance and many experimenters may be faced with problems far beyond the reach thereof and possibly beyond the reach of more advanced texts, some of which have been included in the list of literature. It is not uncommon in dealing with errors of empirical data to be faced with the situation when one's only resort is personal intuition and experience. One should not be reticent in such cases. Individual talent in treating errors backed by expertise acquired through experience is undoubtedly a valuable asset.

There is one golden rule that almost always pays off in experimentation—the rule of patience. He who experiments patiently experiments precisely.

References

1. T. G. BECKWITH and N. L. BUCK. *Mechanical Measurements*, Addison-Wesley, 1961.
2. Y. BEERS. *Introduction to the Theory of Errors.* Addison-Wesley, 1962.
3. H. J. J. BRADDICK. *The Physics of Experimental Methods.* 2nd Edn. Chapman and Hall, 1963.
4. B. J. BRINKWORTH. *An Introduction to Experimentation,* English Universities Press Ltd., 1968.
5. N. H. COOK and E. RABINOWICZ. *Physical Measurement and Analysis.* Addison-Wesley, 1963.
6. M. J. MORONEY. *Fact from Figures.* Penguin Books, 1963.
7. E. M. PUGH and G. H. WINSLOW. *The Analysis of Physical Measurements.* Addison-Wesley, 1961.
8. J. W. RICHARDS. *Interpretation of Technical Data.* Iliffe Books, 1957.

6 *Analysis and Interpretation of Results*

6.1   The requirements

Making the best use of experimental data requires the same ingredients as inventing—99 per cent perspiration and 1 per cent inspiration. Most of the effort has to be put into more or less routine manipulation of the data to put them into a useful form. Then a combination of logic and a questioning mind can enable patterns of behaviour to be detected and understood.

6.2   The virtue of scepticism

As discussed in detail elsewhere, an experimental reading is not to be trusted until its background has been thoroughly investigated. The instrument might lie if it is unsuitable for a particular situation, or if it is in some way faulty. Its presence might interfere with the process in the experiment or it might simply be misread. Any of these mishaps can occur, perhaps even all of them at once. So a good experimenter makes a virtue of adopting a doubting, questioning attitude to his readings.

However, 'moderation is the silken thread running through the pearl chain of all virtues' and obviously the experimenter does not reject all experimental evidence completely. In practice it is found that aeroplanes fly and power stations produce electricity more or less according to plan—and their designs are based on the results of many different experiments. When an experiment does fail badly in its task of providing design information it is likely that the *wrong* experiment has been performed rather than that the readings are inaccurate. For example, in an early attempt to fit a retractable undercarriage to a fighter aeroplane, experiments were undertaken to determine the distortion of the structure caused by aerodynamic forces. Unfortunately the prototype jammed, as the extension caused by the centrifugal forces of the rapidly spinning tyre had been overlooked completely. No amount of care and accuracy in the tests which were undertaken could have compensated for this omission.

6.3 Preliminary work

Before the experiment starts, thought must be given to the best way of dealing with the readings. The likely magnitude of the various measurements should be determined and, if appropriate, a table should be prepared for recording them in an orderly manner.

Plotting readings during the experiment

Instrument readings should normally be plotted as they are taken. Any regions of interest can then be spotted straight away. For maximum value the expected shape of the graph should be sketched out in advance. Then any serious errors in the experimental arrangement will be noticed before it is too late.

If the rough graph is of actual instrument readings without the application of zero errors and scale factors, then there will be no need to perform mental arithmetic in the noise and rush of the laboratory. The need for any corrections should be noted in the log-book so that they are not forgotten—or made twice.

The 'plot as you go' approach was used during a simple experiment which was performed to obtain the characteristic current/voltage curve of a 100 W domestic light bulb. The bulb was wired to an ammeter and voltmeter and connected to a variable voltage source as shown in Fig. 6.1(a). The voltmeter read directly but the ammeter was used with a control marked '× 2' in operation. Before the experiment started a calculation showed that at the rated voltage of 240 V the current should be 0·417 A, which would be indicated by a scale reading of 0·208 on the ammeter. The axes of the graph were prepared accordingly as shown in Fig. 6.1(b). The point R corresponds to the rated output.

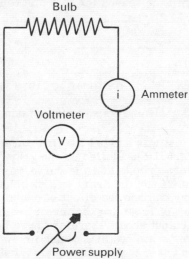

FIG. 6.1(a)   Circuit used when measuring characteristics of an electric light bulb

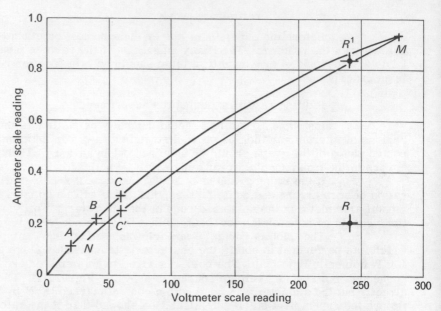

FIG. 6.1(b)   Graph plotted during the light bulb experiment. It became apparent, at point C, that the measured current was considerably different from the expected value (point R). The test was stopped until the reason for this discrepancy had been found

	A	B	C		C'	
Voltmeter scale reading	0	20	40	60		60	
Ammeter scale reading	0	0.115	0.219	0.311		0.248	

FIG. 6.1(c)   Measurements made during the light bulb experiment

When the test was started the first three sets of readings were as shown in the table in Fig. 6.1(c). Plotted on the graph they gave the points A, B and C. It was immediately obvious that the curve was not heading towards point R, so current was switched off and the brain was switched on. It was soon realised that operation of '× 2' control had the effect of doubling the meter deflexion and not, as had been assumed, of requiring the meter reading to be doubled to obtain the current. The rated output should therefore have been represented by the point R^1 (240 V, 0·833 ammeter scale reading) which lay in the track of $A\ B\ C$. Caution is always required when using such ambiguously labelled controls.

What if the readings had just been taken without thought and plotted later? The current for each voltage used would have been taken as twice the meter reading, i.e. four times the correct value. In this simple case the error would probably have been spotted without difficulty. In a more complicated situation where the readings are used in lengthy calculations such errors might pass unnoticed or at least cause unnecessary work.

The graph in Fig. 6.1(b) was made in the laboratory log-book; it was not the one which finally appeared in the report on the experiment as no corrections had been made at that stage for instrument calibration or for the effect of the potential drop across the ammeter.

What about the cases where a reliable point on the curve cannot be obtained so readily? First of all instrument range switches, which are frequently marked ambiguously, should be checked. This can be done by observing the meter reading while the switches are operated under load. It should also be possible to study the instrument manual. Secondly an attempt should always be made to obtain an estimate of reasonable upper and lower limits of each quantity to be measured. For example, in a boiler test most but not all of the heat available in the fuel should go into the steam so any value outside the range of, say, 50 to 95 per cent would be suspect. This sort of accuracy is sufficient to detect gross errors and it can be coupled with an idea of the trend to be expected when measurements are made over a range of conditions. A small proportion of the heat would be wasted near the design output condition and nearly all would be wasted at low steam production rates.

The value of pilot experiments

An experiment has to be preceded by a preliminary trial of the apparatus. The main purpose of this initial test is to check that the equipment is working correctly and to give the experimenter practice in adjusting the controls and reading the instruments. At the same time a set of readings should be taken and the derived results calculated. It can then be seen that the measurements are giving sensible and sufficient information. Provided they lead to satisfactory conclusions, the pilot readings can then be used to provide a check on the readings taken and plotted during the main experiment.

To save everyone's time and to safeguard apparatus from damage during teaching experiments, the pilot tests are frequently performed by the instructor. The students are then advised about instrument ranges and settings, any particular phenomenon to be observed, and the sort of readings to expect. They can thus quickly gain experience of handling the apparatus and of taking readings. However, they must take the trouble to think about the pilot tests and discuss them with the demonstrator. Otherwise they will obtain little value from the experiment and will only perform tasks which are normally delegated to technicians or automatic data-loggers.

6.4   The law of averages

Frequently, when readings are inconsistent, an average value of a number of measurements is calculated. This practice is not always sensible; the use of the mind should precede the use of the mean.

Returning to the light bulb experiment, the plotted results continued up the curve, passing close enough to R^1 for satisfaction and on up to the point M corresponding to the maximum voltage used in the experiment. Current readings were then taken with reducing voltage and plotted giving the curve $MC'N$ as shown. What procedure should have been adopted to obtain the characteristic curve? It would have been easy enough to average the pairs of currents such as C' and C, but this would hide the difference and could lead the experimenter away from some important features of the bulb's behaviour. Instead the table of possibilities shown in Fig. 6.2 was drawn up.

POSSIBILITY	CONSEQUENCE
(a) The current/voltage curve has a hysteresis loop.	Raising the voltage again would cause current values lying on the line OC. The circuit must have unique electrical properties.
(b) The bulb and wiring were hotter at point C' than at point C.	Repeating the test but passing quickly over the region CMC' would reduce the difference.
(c) The bulb was damaged by overloading.	Readings in a new test would lie along the line $ONC'M$. Holding at the maximum voltage would cause a further change.
(d) A fault has developed in one of the instruments.	The test would not be repeatable.
(e) --------- ? ----------	---------------- ? ---------------

FIG. 6.2   Possible explanations for the two curves in Fig. 6.1 (b)

All the possibilities (a) to (d) which the experimenter could visualise were listed together with their consequences so that a series of further experiments could be performed to find the cause of the discrepancy between the two currents. It turned out to be a combination of (b) and (c). The apparatus had heated up *and* the bulb had permanently deteriorated. No original thinking or inspiration was required, but simply a logical, interested approach to the data.

Another situation where simple averages are not meaningful is illustrated by the following example.

In order to determine the frequency of a slowly vibrating structure, the time for 50 complete oscillations was measured. The measured times for the first six attempts were

59, 60, 59, 62, 60 and 60 seconds.

What was the average time? The obvious answer is the total of 360 seconds divided by 6, giving a value 60·0 seconds. But the fourth value (62 seconds) only occurred once and should be scrutinised carefully. Possibilities which suggest themselves are,

(a)   Human error such as miscounting the number of cycles or being slow to operate the clock.

(b)   An isolated fault in the clock.

(c)   A different frequency. The mode of vibration might have changed or a significantly different amplitude might have been used.

The best course in a case like this is to take a few more readings and then apply the James Bond dictum 'Once is happenstance, twice is coincidence, three times is enemy action.' The worst thing to do is to average all the results. An average value is meaningless if the quantities averaged represent different situations. An extreme example of the violation of this rule would be saying that the average beast in a zoo had one wing and three legs.

When it is impossible to take further readings, then the method used in Example (b) on page 137 can be applied to justify the rejection or retention of a dubious reading. In the present case the relevant values are

Standard Deviation $S = 1·095$ s

Deviation with a 50% probability $\epsilon_{MAX} = 1·9$ s

It follows that the fourth reading should be rejected because its deviation is more than 1·88 s

Unthinking calculation of an average value is particularly easy when automatic recording equipment is used to collect a large number of results. The experimenter must supply the imagination which is lacking in the automatic equipment.

6.5   Tables of possibilities

The wise experimenter investigating the light bulb will have made a table of the possible explanations of the observed behaviour. Such a table, normally made in a laboratory log-book, makes a logical investigation possible. Without it, even in this simple experiment, confusion would have arisen. With its aid each possibility was investigated in turn and crossed off when it had been eliminated.

It is always too easy to blame the instruments for any mystery so the possibility of instrument error, (d), was put low on the list. Full consideration was given first to the other possibilities. Instruments inevitably have small calibration errors, occasionally have large calibration errors, but only rarely have the sort of isolated fault which exactly explains strange experimental readings.

When further tests are made to choose from a table of possible explanations for the behaviour of the original experiment, all the items should be checked in order. The tests should not stop as soon as one possibility is found to be operating. Frequently two or more causes act together or in partial opposition.

A most important line is (e)—a blank space waiting to record new ideas. These might arrive unsought but would be essential if all the other items were ruled out.

6.6   Faults in measurements

Before any meaning can be attached to experimental readings, it is necessary to decide
(a) how much the instrumentation interferes with the phenomenon being investigated,
(b) whether the experimental arrangement correctly models the real life situation,
(c) the instrument errors themselves.
These are dealt with in Chapters 4 and 5.

Errors caused by the presence of instruments

It is accepted as a philosophic and thermodynamic truth that, whenever a variable is measured, its value is altered. Ideally the alteration is negligible but in practice it can be very high. An extreme example of intereference occurred when a large aircraft crashed following overheating of a vital oil supply. The overheating was eventually found to be caused by partial blockage of the cooling system by temperature-measuring probes.

On a less disastrous level, difficulties always arise when the rapidly varying pressures in car engine cylinders are measured. The measuring device inevitably increases the cylinder volume and efforts to reduce the volume to normal leave the shape changed.

Since the magnitude of the effect cannot readily be calculated, it must be determined experimentally. One technique is to repeat the tests after deliberately increasing the errors by, for example, fitting a larger, or even a second, probe to the cylinder. The changes in measured behaviour can then be used to estimate the effect of the original probe.

Often this technique can be applied to eliminate the effects of such nuisances as heat losses, friction, and electrical resistance from an experiment. For example, the 'heat transfer coefficient' h between a solid and a fluid is found by heating the outside of a pipe electrically and measuring the temperatures of the tube wall and the fluid within it. If all the electrical energy were to pass into the fluid then h, which is the energy flow per unit area per unit temperature difference, could readily be found. However, some of the electrical energy is lost outside the tube no matter how well the apparatus is insulated. This heat loss will depend on the power supplied. The apparent value of h is plotted against the power supplied as in Fig. 6.3(a). By extrapolating to zero power, when the heat losses would be zero, the true value of h without any heat loss errors is found. Some advice on extrapolation is given in Section 6.10.

Mistakes in the experimental arrangement

Many experiments are quite different from the situations they are supposed to simulate. If the differences are realised and acknowledged then an allowance can be made. Occasionally they are overlooked completely or their possible existence is not considered until the readings are analysed or even until the results have been applied in other work. For example, the arrangement shown in Fig. 6.4 was used to test the resistance of a pipe union to overtightening. In such a union the nut N is tightened to draw the two inner halves A and B of the union together.

Experimental values	A	B	C
Power (W)	50	100	150
Apparent coefficient (W/m^2 °C)	40.43	41.14	41.85

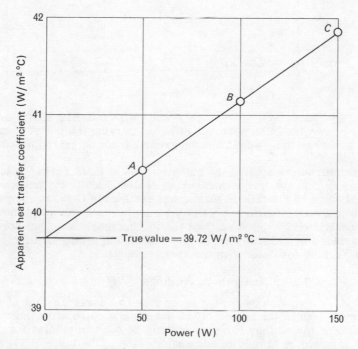

FIG. 6.3(a)   When errors due to losses are unavoidable, a series of measurements under different conditions can be extrapolated to the ideal case of zero loss. A false zero has been used to increase the accuracy

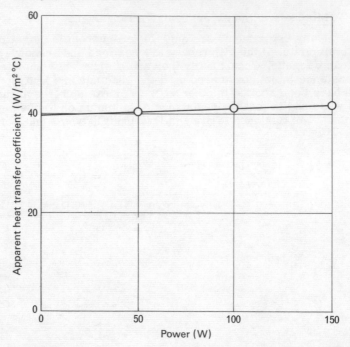

FIG. 6. 3(b)   When the results shown in Fig. 6. 3(a) are plotted without the false zero, the effect of varying the power consumption can be
seen to be small but the graph cannot be extrapolated accurately

If too much torque is applied the nut fractures at point F. In the test the left hand
half A was held in the torque measuring device C while the nut was rotated. The
measured value of the failure torque was far higher than the torque which could be
applied in normal conditions because, during the test, the axial movement of the
nut was restricted by the jaws, C, of the testing machine. This sort of mistake can
only be avoided by carefully observing what is happening in an experimental rig
and by searching for defects in the arrangements.

6.7   Analysis of readings

When an experiment gives rise to a large amount of data, planning is
necessary to reduce the labour of the subsequent calculations. Frequently these
are repetitive and so they can conveniently be done in tabular form. As a trivial
example, it would be simple to add another line to the table in Fig. 6. 1(c) to record
the power if this were required. The tabular form is especially valuable when
each set of readings requires a number of stages of calculation. Comparison of
any intermediate stage of adjacent set of readings is easy and so mistakes can be
detected quickly.

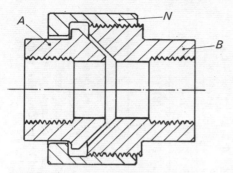

(a) Union before tightening

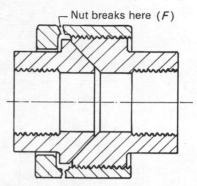

(b) Failure due to overtightening

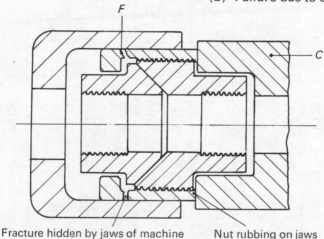

(c) Testing arrangement which gave incorrect failure load

FIG. 6. 4(a), (b), (c)   Testing of a pipe union

For example, the results of a power output and fuel consumption test on a small petrol engine are worked out in the table in Fig. 6.5. The formulae relating the different lines are given. The first four lines are of the measurements and the rest are derived results.

In between stages, as, for example line eight (fuel consumption per hour), are recorded even when they are not required. This need not diminish accuracy. The number can still be retained on the slide rule or calculating machine for calculating the next line.

By glancing along each row there is a good chance of spotting any errors, as the figures should each change smoothly. It can be seen immediately,

	1	2	3	4	5	6	7	8	
1 Speed (rev/min.)	800	900	1 000	1 100	1 200	1 300	1 400	1 500	Experimental observations
2 Brake dead-load (kg)	75·0	75·0	75·0	75·0	75·0	65·0	65·0	65·0	
3 Brake balance reading (kg)	5·8	4·7	5·6	7·1	10·2	3·8	7·9	12·7	
4 Time to burn 50 g of fuel (s)	126	118	109	104	96	91	86	82	
5 Net brake load (kg)	69·2	70·3	69·4	67·9	64·8	61·2	57·1	52·3	(3) − (2)
6 Torque (Nm)	339·4	344·8	340·4	333·0	335·5	300·2	280·1	256·6	(5) × 4·905
7 Power (kW)	4·739	5·416	5·941	6·393	7·027	6·811	6·844	6·715	$(1) \times (6) \times 1\cdot745 \times 10^{-3}$
8 Fuel consumption (g/h)	1 428	1 525	1 651	1 731	1 856	1 978	2 093	2 196	$180 \times 10^3/(4)$
9 Specific fuel consumption (g/kWh)	301	282	278	273	264	290	306	328	(8)/(7)

Full throttle performance test on a small marine engine

Figure 6.5 An example of tabular calculations. The operation required to produce each successive line is noted at the right hand side. Extra significant figures are recorded for all the intermediate stages to prevent rounding off errors accumulating and affecting the final result. By glancing along each row and comparing every column with it neighbouring values, mistakes can be spotted. For example, the 5th value of the torque does not fall between the 4th and the 6th. The mistake was caused by reading the value 64·8 in row 5 as 68·4. This mistake has affected the results in rows 7 and 9.

for instance, that the torque (row 6) in column 5 is incorrect. There remains the possibility of all the calculations being consistently wrong. To avoid this a spot check should be made with one set of values being worked out, preferably by someone else, quite separately.

Accuracy of calculations

Many aids, ranging from tables of logarithms up to computers, are available for processing data. A slide rule is accurate enough in most cases and is especially useful for multiplying a series of numbers by a constant.

More significant figures can be obtained by using a desk calculator or a computer indead of a slide rule but the extra figures are meaningless unless the input data has sufficient accuracy to justify them. No harm is done by using the full capacity of the machine for all stages of the calculation provided sufficient strength of will is available at the end to round off any figures which are put in the report. Unfortunately many people feel that, once a figure has been processed by a computer or calculating machine operating on about a dozen significant figures, any result must be accurate to at least six figures. In fact the accuracy of the calculated results is determined by the accuracy of the input data.

Suppose, for example, that in the timing experiment discussed earlier the frequency were to be calculated. Fifty oscillations took an average time of 59·6 seconds. The frequency is therefore, $50/59·6 = 0·838\,926\,17...\,Hz$. How should this be rounded? Is the frequency 0·8389 Hz? Or 0·839 Hz? It is necessary to reconsider the original readings. These were, again ignoring the maverick 62 seconds,

59, 60, 59, 60 and 60 seconds.

The scale of the stop clock was calibrated in intervals of one second, but it is in order to improve the discrimination by taking a number of measurements. If a single extra time had been measured, it would have been either 59 or 60 seconds, giving the possibilities,

Situation	As measured	New time = 59 seconds	New time = 60 seconds
Average Time (s)	59·6 (exactly)	59·5 (exactly)	59·6
Frequency (Hz)	0·838 926 17, . . .	0·840 336 13. . .	0·837 988 8.

It can be seen that the third figure after the decimal place changes considerably; the correct frequency is no more precise than

0·84 Hz

Even this figure cannot be claimed to be accurate, since no account has been taken of the accuracy of the clock. There is no need to construct a table every time a reading is taken, it was included here merely to emphasize the point.

Many modern measuring systems provide a digital output either directly or into a data logger. Here again care is required to differentiate between precision of the measurement and number of significant figures provided by the system. Information has to be gleaned from observation of the steadiness and repeatability of a measurement and from a knowledge of the type of transducers used.

Intermediate steps in calculations

For clarity in the example of timing given above the average time was calculated for each of the three cases considered. Normally a better approach is to avoid any intermediate multiplication and division with its consequent 'rounding off' error and to say (considering the figures on the centre column)

Total time for 6×50 oscillations =

$$59 + 60 + 59 + 60 + 60 + 59 = 357 \text{ secs.}$$

$$\text{Average frequency} = 6 \times 50/357 = 0 \cdot 840\,336\,13 \text{ Hz}$$

$$= 0 \cdot 84 \text{ Hz}$$

Of course, in this particular example the time could have been recorded for 300 oscillations in the first place but then no check on consistency would have been available and considerable concentration would have been needed to count the 300 cycles without a break.

It should be noticed that the units of every quantity are included at each stage. Thus the derived units can then be checked and the risk of accidently inverting a division is reduced.

Preservation of calculations

All calculations and deductions should be written in either the laboratory log-book or a special calculation book. They need not be reproduced in full in the laboratory report but must be available both to answer any subsequent queries and as a model for dealing with similar work. Instead of relying on memory, all assumptions, symbols, units and other relevant information should be clearly explain alongside the calculations.

6.8   Graphical analysis

Frequently experimental data can be analysed graphically with a saving in labour and time compared with numerical analysis. An incidental advantage of graphical methods is that they give a clearer picture of trends in behaviour and of scatter in experimental results. Indeed, when organising a computer program to handle the data from a long series of experiments, it is sensible to analyse the first few sets of results graphically to make sure that no special features of the results disappear for ever inside the computer.

Points can be plotted on a graph with the same precision as most instruments give. If a fine line is drawn as a good fit to the points it can average out scatter and so improve on the accuracy of individual readings. Any points which do not fit into the general pattern can be spotted. After investigation they can be studied further or ignored as appropriate.

Mistakes to be avoided

The potential accuracy of graphical analysis can be destroyed by two common mistakes when drawing the curves. These are illustrated in the examples described below. In each case the scales were well chosen and the individual points plotted accurately but the curve drawn was not the best for the plotted data. Further advice on drawing graphs is given in Ch. 7, 'Graphs', p. 123.

The magnetic origin. Figure 6.6(a) shows the results of measurements of force and deflexion made when finding the stiffness of a spring. When the graph was drawn it seemed reasonable that, as zero deflexion required zero force, the origin was the most accurate point on the graph. Therefore the graph was drawn from the origin through the plotted points with about the same number on each side. Unfortunately the common, and natural, assumption about the importance of the origin was wrong. At the best the origin is just another measured point. Frequently, as in this case, it is the least reliable measurement of the lot. Measuring instruments and transducers are likely to have a certain amount of 'backlash', i.e. clearance between the moving parts which is taken up before any

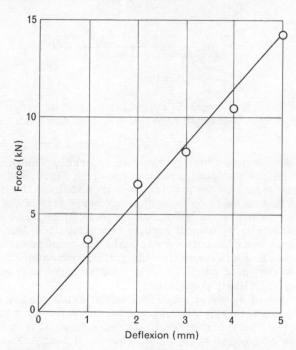

FIG. 6.6(a)   The magnetic origin mistake. The graph as orginally drawn. It passes through the origin but is a poor fit to the remaining points

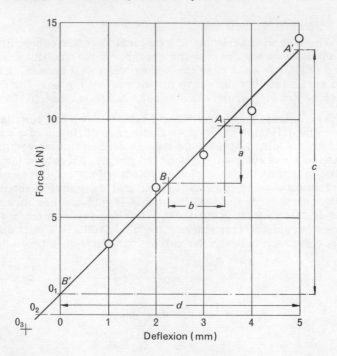

FIG. 6. 6(b)   The magnetic origin mistake. The graph redrawn as the best straight line through the plotted points and ignoring the origin

output is displayed. Subsequent readings are then in error. The best straight line through the data, but with the origin ignored, is shown in Fig. 6. 6(b).

Since the nature of the problem requires that the line must pass through the origin, it is necessary to move the reference grid of the graph until the origin lies on the line. Whether the origin should be at 0_1 or 0_2 or at some other point cannot be decided reliably without further experiment. In this case no more information is available so it is necessary to make the most reasonable choice. If 0_1 were to be chosen, it would imply that the deflexion measurements were correct but that the experiment started with a negative force applied to the spring. This does not seem a very likely possibility.

The choice of 0_2 would mean that no deflexion reading occurred until a certain amount of clearance was taken up and all the deflexion readings should be increased by $0_2 0$ (0·42 mm) above the measured values. It is quite likely, however, that the true origin lies at an undefinable point such as 0_3.

Zero errors are just as likely to be found on non-linear graphs. To avoid distorting the curve the origin should be ignored (or at least not given special emphasis) when the graph is drawn.

The critical point mistake. When an experimenter is plotting a set of his readings, he is often reluctant to acknowledge that features of importance can lie between his measured points. He then draws his curve so that any point of inflexion, maximum or minimum, coincides exactly with a plotted point. An example is shown in Fig. 6.7(a). Here the variable y rises to a maximum value then decreases as variable x increases. All too often the curve would be drawn as shown. A better curve would have the maximum lying between two of the plotted points as shown in Fig. 6.7(b).

The mistake is partly caused by a reluctance to draw the graph through values which are higher than any of the readings taken. It is helpful if the maximum value of an independent variable such as y is recorded. A horizontal line can then be drawn at the maximum value to form a tangent to the curve as in Fig. 6.7(c). Unfortunately this stratagem cannot be applied in many cases. For example, strains might be measured at a number of fixed points on a structure without any information being available between these points.

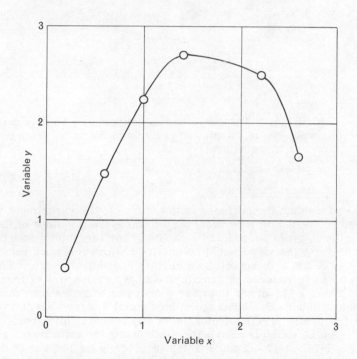

FIG. 6.7(a)   The 'critical point' mistake. The curve has been drawn through all the plotted points but there was no justification for assuming that the maximum value occurred exactly at a measured point

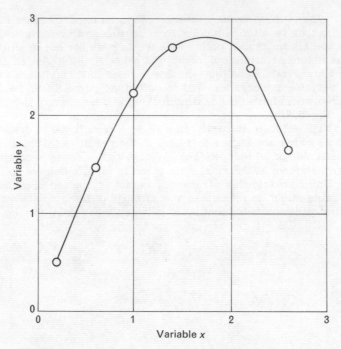

FIG. 6. 7(b)   The curve drawn correctly. The maximum occurs between two of the plotted points and the curve is smoother and appears more natural

Measurement of slope

To obtain the stiffness of the spring in the experiment whose results were plotted in Fig. 6. 6(b) it is necessary to measure the slope of the graph. Mathematically it would seem that by choosing any two points, such as A and B, and calculating a/b the slope would be obtained. However, when reading off the coordinates of A and B there will be a certain unavoidable error. The significance of this error can be reduced by spacing A and B as far apart as reasonable at, for example A' and B'. The error in either a or b can be reduced by making it a convenient whole number. By choosing the horizontal distance the division can also be simplified, frequently with the avoidance of any rounding off error.

In the example shown, the slope using the coordinates of A and B is $a/b = (9\cdot65 - 6\cdot72)/(3\cdot44 - 2\cdot25) = 2\cdot462\ 18\ldots$ kN/mm. The use of the points A' and B' gives the more accurate value $c/d = (13\cdot57 - 1\cdot08)/(5\cdot00 - 0\cdot00) = 2\cdot498$ kN/mm which is close to the value of $2\cdot496$ kN/mm calculated by the method of least squares described in Ch. 5. As originally drawn, the slope was $14\cdot17/5\cdot00 = 2\cdot834$ kN/mm.

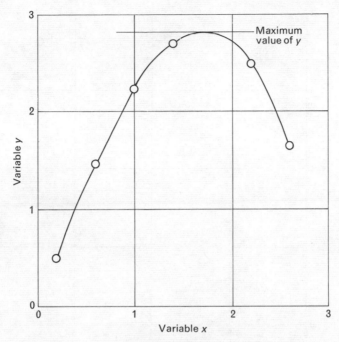

FIG. 6. 7(c)   It is of assistance when plotting graphs if the maximum value of a variable has been noted

When taking measurements from graphs, it is best to use the printed scale and not a rule. This is because the printing process and subsequent paper shrinkage can cause considerable variations from the nominal size. Furthermore these effects are directional; very few of the 'squares' on graph paper are, in practice, square.

Fitting curves to experimental data

It is not easy to draw the 'best' curve through a series of points such as those shown in Fig. 6. 8(a). A choice has to be made between four possible assessments of the situation. This choice must be a conscious one and the correct assessment will depend on the source of the data as well as the arrangement of the plotted points.

A simple curve can be drawn by eye or with the help of numerical curve-fitting techniques. For example, the best fitting parabola is drawn in Fig. 6. 8(b). This particular curve can only be correct if the measured values of X and Y are liable to have the errors which can now be seen.

If the expected errors are less than those shown in Fig. 6. 8(b) then the obvious step is to draw a more complicated curve. The cubic in Fig. 6. 8(c)

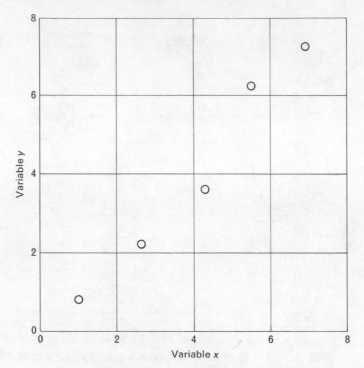

FIG. 6. 8(a)   Experimental data with an appreciable amount of scatter.
There is no single 'best' curve through a set of points

involves smaller erors but can only be correct if the relationship between X and
Y just happens to have a point of inflexion in the middle of the region of interest
and a maximum and minimum just outside the ends. By drawing a higher order
polynomial or Fourier curve the plotted points can be fitted exactly. It is most
unlikely that such a curve would be the correct one, however. The test to apply
before choosing a complicated curve is to remove one or more points. Since the
original choice of values of X will have been arbitrary, the omission of one point
should not make any real difference. If the proposed shape now looks unnecessar-
ily complicated it is probably wrong. For example, if no reading had been available for $X = 5 \cdot 50$ (the fourth point) then the upward bulge in the right-hand part
of the curve would have no purpose. The complexity of the curve in Fig. 6. 8(c) is
only required for this one point. This is not to say such a curve cannot be correct.
There is not sufficient evidence to make a decision without further points in the
regions of $X = 5$ and $X = 6$.
          The next possibility to consider is that, rather than most of the
points having large errors, just one point contains a serious mistake. If, for in-

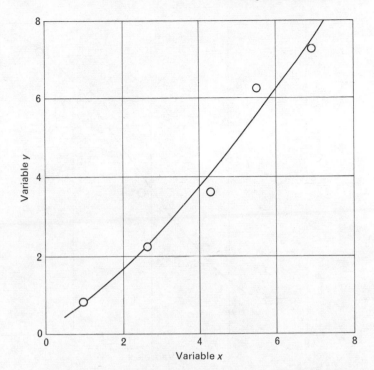

FIG. 6. 8(b)   This simple smooth curve can only be correct if the large errors in the third and fourth points are likely to occur

stance, the fourth point is discarded, a simple parabolic curve can be drawn (Fig. 6. 8(d)) to give a good fit with the remaining points.

The final possible explanation for the 'scatter' is that an instrument fault could have started between the third and fourth readings. This could have been caused, for example, by accidental interference with a zeroing device or by changing the range setting on an incorrectly calibrated instrument. In such a case the best curve would be similar to that in Fig. 6. 8(e)—a discontinuous line.

Points cannot be lightly discarded whenever they fail to fit preconceived notions of the shape of the graph or just because they make the graph look untidy. They must first be carefully checked and the source of the mistake or the unusually large error found. This is made easier by the recommended practice of plotting results before finishing the experiment. Drawing a curve through scattered data is in many ways similar to the process of finding the average of a number of readings which was discussed earlier.

The above difficulties can sometimes be overcome by taking a large number of experimental readings but this is expensive and time consuming. A

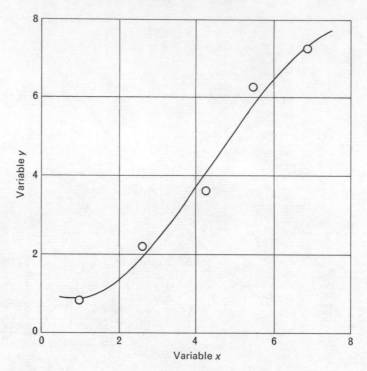

FIG. 6. 8(c)   A more complicated curve is a closer fit but would hardly be justified by the limited data

better approach is to plot the data to scales chosen so that a straight line is form-ed. An explained in Chapter 2, some trial and error work is required but an in-telligent approach can reduce this to the minimum. Advantage should be taken of any possibility of qualitative reasoning about extreme values, particularly when only a small quantity of experimental data is available.

For example, a number of experiments were made to find out how the different dimensions and the choice of materials affected the deflexion of a certain type of structure under the action of a certain load. Because only two materials were available only two results could be obtained when the effect of Young's Modulus was being investigated. These were plotted as shown in Fig. 6. 9(a). It might be thought that the only graph which could be drawn would be the straight line C D as shown. However, this would have implied that, for a certain modulus D, the deflexion would always be zero and for a greater modulus the de-flexion would always be negative. At the other extreme, the use of a material with no resistance to deformation, that is with a Young's Modulus of zero, would cause

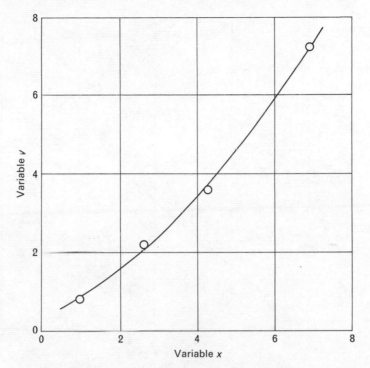

FIG. 6. 8(d)   If it is accepted that a large error (or perhaps a mis-take) could affect one point, a simple curve can fit the remaining points closely

the structure to have a finite deflexion $O\,C = 4\cdot74$ mm. It was clear that the straight line, or indeed any curve cutting the axes, would not satisfy the physical realities of the structure.

It was decided that some sort of hyperbola was required. The results were plotted again, this time with $1/E$ as the independent variable. The two mea-surements were found to lie on a straight line through the origin as shown in Fig. 6. 9(b). The origin corresponds to an infinite value for Young's modulus, i.e. to a completely rigid material, and to zero deflexion which is a logical combina-tion although physically unattainable. Therefore it was decided that the straight line was correct. The conclusion was drawn that the deflexion was inversely pro-portional to Young's Modulus, a result which could, in fact, have been ascertained from dimensional analysis in this case. By measuring the slope of the line, the formula deflexion (mm) $= 246\cdot5/$(Young's Modulus (GN/m^2)) was obtained. This equation was used to enable sufficient points to be plotted for the graph in Fig. 6. 9(c) to be drawn.

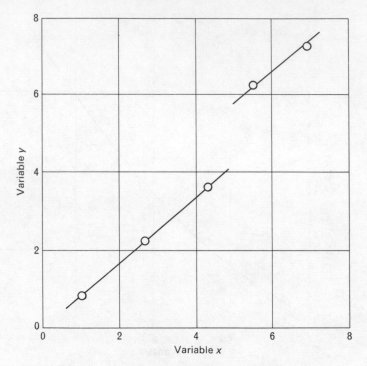

FIG. 6. 8(e)   Yet another possibility for the best fit. A change in the zero error between readings could break the curve into two disconnected parts.

Similar physical reasoning to that used above has been applied on occasion to a situation where only a single experimental point was available but this is not a practice to be recommended. Even the two points in this example might not be thought sufficient.

Log-log graphs

Suppose that in the previous example the graph of deflexion against 1/Young's Modulus had not been a straight line through the origin. It would have been possible, but tedious, to search for a solution by trying 1/(Young's Modulus)^2, 1/(Young's Modulus)^3 and so on. Sometimes dimensional analysis can be used to select the correct power—but it cannot help if, for example, the variables are dimensionless numbers.

When it is suspected that two variables x and y are related by a power law of the form

$$y = ax^n$$

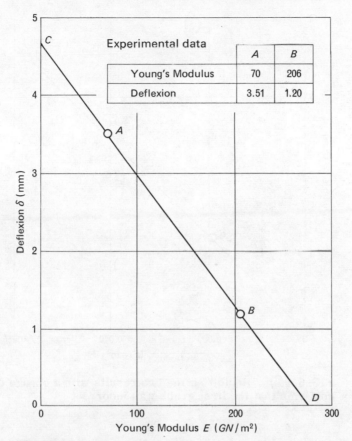

FIG. 6. 9(a)   Drawing a graph from limited data. A straight line through the plotted points A and B seemed the simplest form of curve but it leads to impossible situations if extended to C and D

then, taking logarithms of both sides,

$$\log y = \log a + n \log x.$$

In other words, a graph of $\log y$ against $\log x$ will be a straight line of slope n (n can be positive or negative). To avoid having to look up a series of logarithms it is normal to use paper with logarithmic scales on which the lines of the reference grid are spaced in a similar manner to markings on a slide-rule scale.

An example of the use of such a plotting technique occurred during the tests on the structure described earlier. Four structures each differing only

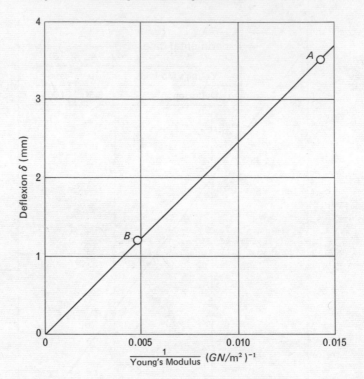

FIG. 6.9(b)   Replotting the two results with a change of variable confirms that the first graph was incorrect

in the one dimension, D, were tested with a single value of load and the deflexion was measured in each case. The corresponding dimensions and deflexions were plotted as in Fig. 6.10(a). The four points were then plotted on a logarithmic scale as shown in Fig. 6.10(b). They were found to lie on a straight line whose slope was measured as $-a/b = -2\cdot93$. Since most engineering systems are found to have simple values for exponents in governing equations, the value of n was taken as being exactly minus three. The data were plotted once more but this time as a graph of deflexion against $1/D^3$ (see Fig. 6.10(c)) and were found to lie on a straight line through the origin indicating that

$$\delta = C/D^3$$

The value of the constant, C, was found to be 11330 mm^4 by measuring the slope of the graph. It could, of course, have been obtained directly from the log-log graph but it would have been difficult to read the scales sufficiently accurately.

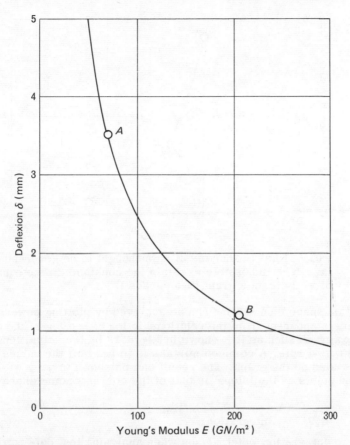

FIG. 6. 9(c)   The correct shape is a rectangular hyperbola as the deflexion is inversely proportional to the modulus

Paper with logarithmic scales has many advantages in situations such as this but it can be infuriatingly confusing until its properties are familiar. It is necessary to have available a number of styles of paper to cover different ranges of values. The number of 'decades' of scale needed in each direction depends on the ratio of the greatest to the least value. In the previous example the deflexion covered a range of 90·5 down to 3·36 mm, i.e. there is a ratio of about 100 : 1 between the nearest round numbers so two 'decades' of paper are required. The normal rules of scaling the variables do not apply; the physical length of the graph depends on the size of one decade on the printed paper. An attempt to double or halve the values of deflexion to alter the size would merely move the line up or down without altering its slope or length. If it is important to fit a log-log graph

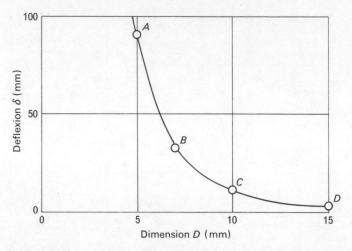

FIG. 6.10(a)   Analysis of data thought to be governed by a power law. It is impossible to obtain the constants in the equation $\delta = AD^n$ directly from a graph of δ versus D

into a certain space then scaling can be achieved by plotting powers of the variables. This is apparent from the relationship $\log x^p = p \log x$. To obtain the slope of a log-log graph such as that shown in Fig. 6.10(b), the actual lengths a and b are measured with a rule. A common mistake is to read off the values of the variables from the scales on the graph. The result can be shown to vary with the positions of the two end points as the 'slope' is that of the corresponding chord to the original curve.

6.9   Integration

Integration which arises when analysing test data can be performed graphically, numerically or, occasionally, analytically. For example, in order to find the volume of oil flowing through a circular pipe, a probe was introduced which measured the velocity. As the skin friction causes the fluid velocity to vary from zero at the surface to a maximum at the centre, measurements of the velocity v were made at each of a number of radii, r. The readings are shown in Fig. 6.11(a). By considering the volume flowing in an annular element of radius r and thickness δr the total volume flow can be shown to be given by

$$Q = \int_{0}^{r_0} 2\pi r v \, dr$$

where r_0 is the radius of the pipe which was 25 mm in the case considered. The values of rv were calculated and plotted as shown in Fig. 6.11(b). The volume flowing per second, which is equal to 2π multiplied by the area under this second graph can now be obtained in a number of different ways.

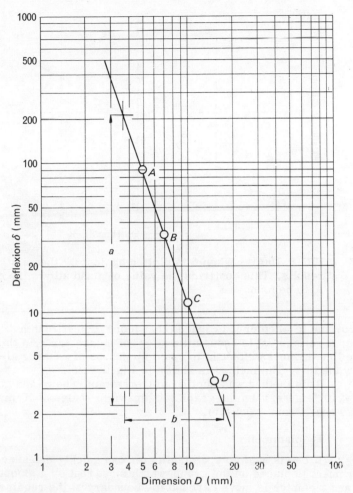

FIG. 6.10(b)   Replotted to logarithmic scales the graph becomes a straight line whose slope is equal to the constant n (in this case n = −3)

By counting squares

This elementary method should not be dismissed as childish. The technique is to count the number of squares of the reference grid lying beneath the curve. Most can be disposed of in blocks of 100 or 10 000, leaving those near the curve to be counted individually. Squares broken by the line are rounded up or

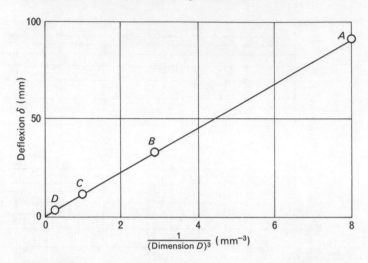

FIG. 6.10(c)   When δ is plotted against D^n then a straight line again results. This confirms the value of n and allows the other constant, A, to be found

down according to whether it is judged that more or less than half lies beneath the curve. A refinement to reduce the effect of human error in this judgement and to reduce line thickness effect is to count up groups of squares alternatively above and below the line. The method is shown in Fig. 6.11(c).

This method was applied to the example shown and gave a volume flow of 0.615×10^{-3} m³/s when the graph was drawn to scales of 10 mm ≡ 0.5×10^{-3} m²/5 and 10 mm ≡ 0.4 mm.

By planimeter

If one is available, a planimeter can quickly measure irregular areas. The instrument is placed on the graph, which must rest on a smooth horizontal surface, and a pointer is moved around the boundary of the required area. The reading on the instrument is then multiplied by a scaling constant. The constant should be determined by measuring a known rectangular area of the graph paper as this will automatically compensate for paper shrinkage. When a planimeter was used on the graph in Fig. 6.11(b), a value of 0.621×10^{-3} m³/s was obtained for the flow rate.

By numerical calculation

Several formulae such as the trapezoidal rule and various forms of Simpson's rule are available. They should not normally be applied directly to measured data as they cannot make any allowances for scatter and the occasional

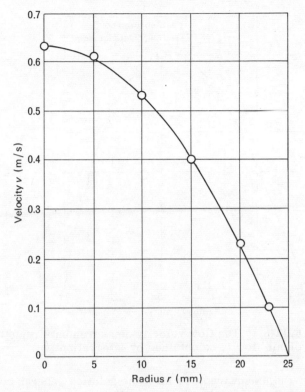

FIG. 6.11(a)   The variation of fluid velocity with radius in a circular pipe. An 'average' velocity would be needed to obtain the volume flowing per second

misreading of an instrument. When applied to the pipe flow problem Simpson's rule yielded a volume flow of 0.624×10^{-3} m³/s.

In an attempt to improve the accuracy, the calculation was repeated using additional points interpolated from the graph. A new value of 0.615×10^{-3} m³/s was obtained.

By curve fitting

If an algebraic expression can be found to fit the experimental data, an algebraic expression for the integral will normally follow. Not only does this give a quick value for the integral but the solution is then available to be applied to integrate other, similar, sets of data with very little extra work.

In the particular case being discussed, it appeared that the 'velocity profile' was parabolic. Accordingly the graph shown in Fig. 6.11(d) was plotted

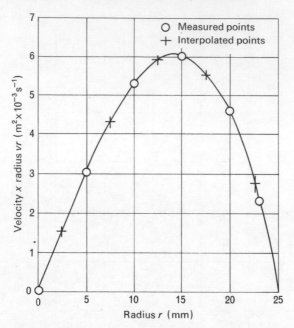

FIG. 6.11(b)   The flow velocity measurements replotted to enable
the volume flow to be obtained by integration

of velocity against radius squared. To keep the solution general, both the variables
were rendered non-dimensional by dividing by their respective maximum values.
The straight line graph shown was fitted by the equation

$$V/V_{max} = 1 - (r/r_0)^2$$

from this expression the volume flow was calculated to be equal to $\pi v_{max} r_0{}^2/2$
which equalled $0 \cdot 618 \times 10^{-3}$ m^3/s in the case under consideration.

6.10   Extrapolation and interpolation

It is often found that measurements cannot be made with the apparatus
operating at the exact conditions for which the results are sought. This might be
because the variable which should be set to a particular value has to be regarded
as a dependent variable for the purpose of the experiment. For example, to find
the position of a machine part at a given time would be difficult, while it might be
easy to measure the time at which it reaches a series of fixed positions. Again
the independent variable might not be infinitely adjustable—a force might be
applied by means of dead-weights which increase by inconveniently large increments. A third possibility is that in the experiment the range of values available

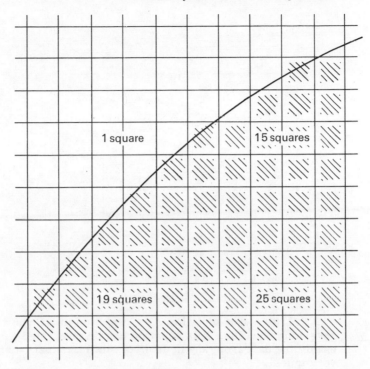

FIG. 6. 11(c)   Obtaining the area under a curve by 'counting squares'.
Each square shown shaded counts as one complete unit. The unmarked
squares count as zero. Human error can be reduced by counting
alternate blocks above and below the line

for a variable might be too small. For example, decisions involving the long term
strength or corrosion resistance of a material might have to be based on a relatively short test.

In all these cases the conditions at the required point can be read
from a graph drawn through the available data. When the required point lies inside
the range of plotted values the process is known as *interpolation*, otherwise it is
extrapolation. Of course there is no need to draw a graph in every case, mathematical methods, e.g. curve fitting, can be used.

Interpolation

Probably the easiest method is to draw a graph. The scales should
be chosen to give a fairly flat curve or, better still, a straight line. A false origin
might improve accuracy as in the example illustrated in Fig. 6. 3. As always, the

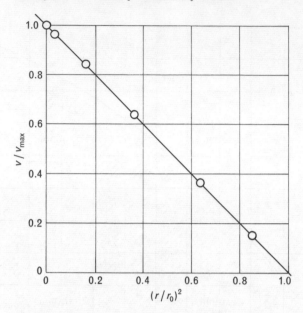

FIG. 6.11(d)   The pipe flow data replotted in a form which gives the governing equation. This equation can be integrated analytically to enable the data from any number of similar situations to be handled very quickly

common mistakes of graph drawing which were discussed previously must be avoided.

It is just possible that an unfortunate change in the character of the curve might happen between the plotted points. In Fig. 6.12(a) the measured amplitude of vibration of a motor car body at four particular frequencies is plotted. From the curve drawn through the experimental data the amplitude corresponding to a frequency of 2·5 Hz appears to be 6 mm. The real value at this frequency is nearly four times greater (21 mm) as can be seen from Fig. 6.12(b), where the same data has been plotted with additional values at other frequencies. This phenomenon (resonance) is well known in both electrical and mechanical vibrations and the experimenter would normally be on the look out for it. In less well known cases similar behaviour can be detected by careful observation of the instruments and the apparatus while the variables are being changed. Where possible the measured values should be taken close to the interpolated points.

Extrapolation

Extrapolation is always less certain. It inevitably involves venturing into unexplored regions, so extreme caution is necessary. Consider, for example, the measurements illustrated in Fig. 6.12(c). Three pairs of values of the variables M and m lie on a straight line A B C. It seems reasonable to extrapolate this line a small way to the point D. The question is, what is the limit of such extrapolation? Is it also reasonable to extrapolate to point E and beyond? It is necessary once again to consider the physical situation. In this case M is the mass of a certain baby in kilogrammes and m is his age in months. Experience shows that in similar situations the mass does not increase linearly with time for long. It is unlikely that the child would 'weigh' 27 kg (60 lbm) at the age of four years as represented by point E and unthinkable that after 20 years ($m = 240$) he would 'weigh' 123 kg (272 lbm). Fortunately it is not necessary to have experience of

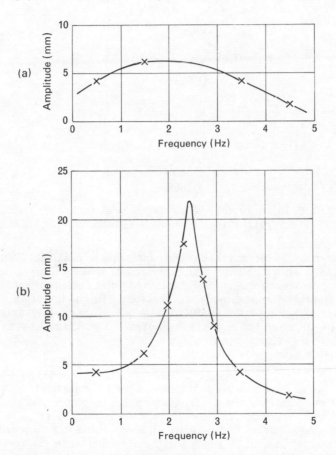

FIG. 6.12   An illustration of the difficulty of interpolation. The first curve looks right but the additional data used in the lower graph leads to an entirely different shape.

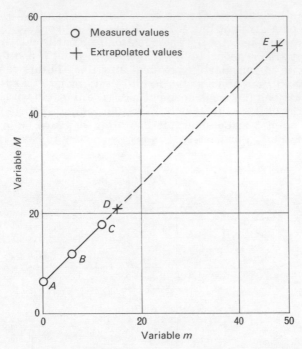

FIG. 6.12(a),(b),(c)   Extrapolation of data. The points A, B and C
lie on a straight line. Is point D reasonable? and E?

similar cases to avoid the mistake of extrapolating to point E. There is a golden
rule of extrapolation which says that 'Extrapolation of a curve is not to be trusted
when it is interpolation between an established point and the point infinity-infinity'.
In a real physical system the mode of behaviour will change at an unknown point
along the curve. Extrapolation right up to the point where one variable (say x) is
infinite is quite in order but is difficult to draw. It is better to rearrange the
variable as $1/x$ and extrapolate to zero.

6.11   Inspiration

So far only routine analysis has been considered. This is sufficient
for nearly all experiments and the techniques can be tested and improved by practice. For a small proportion of experiments, something more is required. This
extra is much more elusive and is difficult, some would say impossible, to teach.
Every experimenter should be ready for the occasion when the results fall into a
new pattern. It might be that experimental scatter is seen to be the effect of a
previously ignored variable or perhaps a quite simple relationship can be spotted
between two variables. Although little is known of the reason why sometimes

'things click into place', it is known that waiting patiently for them to do so is not sufficient. Waiting, for a day or so, can help but only after a problem has been considered and worried over in detail for long enough for the subconscious mind to take hold of it. Then stimuli from other problems or everyday living can be enough to give a new insight or a different viewpoint. For example, a highly success-ful altitude-measuring device for aircraft is reputed to have been inspired by a pair of spotlights in a theatre. The inspiration only came because the inventor had already spent a lot of effort seeking a way of measuring altitude before relaxing.

The subconscious mind can be helped in its task if a conscious effort is made to look at a problem differently. A situation can be described accurately but differently by changing the viewpoint. A pessimist's half empty wine bottle will appear half full to the optimist. One person's careful judgment will seem guesswork to another person. It is sometimes useful to look at a situation from the directly opposite point of view to the obvious one. During the search for a suitable material for the light-sensitive drum in the Xerox photo-copier, selenium was tried and rejected as insufficiently sensitive. Then it was realised that it was *too* sensitive for the experiment—the stray light had 'fogged out' the image. A 'snag' which spoils a reading could be regarded as a fortuitous demonstration of a par-ticular effect instead of a confounded nuisance. A spot of mould which spoiled a cul-ture of bacteria led Fleming to the discovery of penicillin.

7 *Communication*

7.1   The need for effective communication

When an experiment, calculation or other task is completed and the significance of the results has been understood it is difficult for the engineer to maintain his enthusiasm. But it should be remembered that his work will be largely wasted until two final stages are accomplished. First the information and experience gained must be assembled into a report of some form and then potential users of the information must read or hear and understand the report.

These last two stages deserve, but often fail to get, the same careful attention as the other stages of the work.

On other occasions it will be necessary to explain new ideas or to persuade someone to agree to particular proposals. In fact, from his first job interview right through his career to his retirement speech, an engineer will spend more of his time communicating than he spends calculating. He should be prepared, during his training and afterwards, to work hard increasing his effectiveness in this aspect of his work.

Although only a few people have a natural flair for communication, anyone can achieve a high degree of competence with practice and by always thinking in advance of the effect that each sentence, paragraph and illustration will have on his reader or audience.

To be successful, a communication has to hold the attention of the reader or listener against a host of unavoidable distractions. It must be kept as short and as interesting as possible, while conveying the desired information clearly and completely.

7.2   Preparing a written report

Written reports require different techniques from lectures, but the preliminary work is the same in each case. Producing a report should never be regarded as an end in itself: every report should have a definite purpose which must be defined before work starts. This purpose *has to* be allowed to govern the form the report takes. Many bad reports are produced because an attempt is made to mould them into a standard form laid down by a particular firm or school. The quality of a report should be measured by how well it does its job of transmitting information and not how closely it fits some arbitrary formula such as 'a picture, two tables, three graphs and 750 words'.

The recipient

When preparing a report or lecture, it is vital to visualise the attitudes and the competence of the people who will receive it. Reports on a single topic appearing in *The Engineer* and in *The Guardian* should be quite different even if they are written by the same person. A student's descriptions of his motor cycle should be different for his colleagues and for his girl-friend. Different people have different interests and different previous experiences and these should be borne in mind by a writer or speaker.

Sometimes a single report must serve more than one group of people and here special care is needed or none will find the report satisfactory. Normally the interests of one group have to be given priority but the others cannot be completely neglected.

Student laboratory reports are a special case. Although read and marked by someone with extensive experience of the relevant subject, they should be written as though intended for a reader at about the same level of technical training as the student. Another peculiarity of such reports is that frequently it is the preliminary work only which is submitted for marking—the final report never being produced.

It is sometimes difficult, after working on a problem, to appreciate which points will appear difficult to someone reading or listening to a description of it for the first time. Nearly every student has experience of a lesson which seemed completely obscure because the lecturer had not realised the necessity of explaining an 'obvious' detail or giving some 'well-known' background information.

7.3 Guidance on the arrangement of written reports

Certain conventions must always be observed but these leave room for individual variations in style of presentation. The subject matter should be arranged in the order shown below. This familiar, logical order enables a reader to find and inspect a particular item without working through the whole report. The complete list shown is suitable for reporting on investigations lasting many months. For less ambitious projects including student laboratory experiments, some simplification is necessary.

While many examples of this layout can be found in the proceedings of learned societies, it should be realised that the standard of writing is not uniformly high.

The normal layout

¶ 1. *Title and author's name.* The title should be short but should convey a clear idea of the subject matter. For example, *Power Output Measurements on a BLMC Series E Petrol Engine* rather than just *Petrol Engine*.

¶ 2. *Summary*. This section briefly outlines the purpose and extent of the work reported and gives the main conclusions reached. The purpose should be introduced as a report on the work accomplished rather than a bald statement of a target. For example, it is better to say 'measurements were made of the thermal conductivity of PVC' than 'Object: to measure the thermal conductivity of PVC'. This is because, by the time the report is being written, the aim of the experiment has already been achieved. The summary is designed to enable a reader to decide quickly whether the report contains anything valuable in his field of interest. It should be worded to help him, not to trap him into wasting time on a report which is of no value to him.

There is no need to give this paragraph the title 'Summary'. Its position between the main title and the body of the report is sufficient to identify its purpose.

¶ 3. *Contents*. A list of contents, giving the page number of each section or sub-section and each illustration, is required in long reports. One should be included *only* when it is expected to be of real assistance to the reader and not just to make a report seem more formal.

¶ 4. *Notation*. Next comes a list briefly defining the meaning, and units where appropriate, of every symbol appearing in the report. The symbols are arranged in alphabetical order, English letters first, followed by Greek. Some authors also define each symbol when it first appears in the body of the report but it is normally sufficient to do this for unusual symbols only.

Some recommended symbols are shown in Table 2 of Appendix 3. A more comprehensive list is given in BS 1991 [1].

¶ 5. *Introduction*. The introduction should put the work reported into perspective in the relevant field of interest. It should start with a brief review of previous work in the same topic giving references to existing publications and then continue with a justification for doing the work to be reported. Finally the range of the current work should be outlined.

¶ 6. *Theory*. The relevant theory, if there is any, should be given, but in outline only. Each basic equation should be introduced with an explanation, where appropriate, of its physical significance. With a brief description of the method of manipulation, but without giving detailed intermediate steps, any derived equations or solutions should then be written. Sometimes, even in outline form, the theory will be excessively long. In this case, or when a lot of detail is required to explain an unusual method, it is sensible to give just the theoretical conclusions at this stage and to put the full theory in an appendix at the end of the report. Once more the author must bear in mind his typical reader's capability and experience. He must also help a newcomer to the particular topic by giving references to the previously published theories on to which his own work was built.

¶ 7. *Experimental arrangements*. The apparatus and experimental procedure should be described, any detailed descriptions of novel or complicated equipment being relegated to a further appendix. Carefully prepared line drawings, and very occasionally photographs, should be included where they will help the reader to understand the arrangement. A report should not normally contain a list of every item of equipment used or of identification numbers of measuring equipment. This information, which is of course recorded in a laboratory log-book, is of no interest to the normal reader.

¶ 8. *Results.* The results obtained from both theory and experiment should be given next. Frequently a graphical presentation is the easiest to understand and enables the differences between theory and experiment to be seen more quickly. It is not normally necessary to give results in *both* numerical and graphical forms, but where an expensive or complicated experiment has been performed to obtain basic physical properties, the actual readings taken should be included in the report. Experimental measurements will anyway be preserved indefinitely in the laboratory log-book.

The units must be given for all physical quantities. The units used should be those normally accepted and therefore familiar to the reader. They should be consistent throughout the report. For example, if temperatures are quoted in degrees Fahrenheit then enthalpy should be in British Thermal Units. Where a report is written for readers who use both Imperial and SI units, then one system should be chosen as the main one but values in the other system should be added in brackets, e.g. '... a mass of 0·97 kg (2·14 lbm)'. As well as measurements of physical quantities this section should contain descriptive results, with photographs and line drawings where helpful, of such items as flow behaviour and fracture surfaces.

¶ 9. *Discussion and suggestions for further work.* The author's own comments on his work appear here. He might wish to point out strengths and weaknesses of his experimental arrangement or of his method of calculation. He should compare his results with those of other workers. Since all experimental and theoretical investigations are part of a continuous unfolding of new knowledge each report should illuminate the way a little further. The next stage of development should be suggested here or, occasionally, it can be stated that a particular line of enquiry is no longer worth pursuing. Further work which the author has already started should be mentioned to reduce the chances of unnecessary duplication.

¶ 10. *Conclusions.* In this section the main results are summarised and their significance is briefly explained. For example, there might be a good or bad agreement between experiment and a given theory. An important possibility to bear in mind is that a set of experiments might give inconclusive results not capable of either confirming or disproving a particular theory. Outside schools and universities, few reports are concerned with confirming theories. The conclusions are more likely to be about the suitability of a given design for a particular purpose or of a new material for a certain duty. This section should be kept short enough for the main conclusions to be quickly found by a casual reader.

¶ 11. *Acknowledgements.* As a matter of courtesy, thanks should be expressed for any assistance in the form of facilities or advice. This custom has the advantage of making known departments and people who have developed particular skills in auxiliary fields as well as giving credit where it is due.

¶ 12. *Appendices.* As already mentioned, detailed theory and descriptions of apparatus should be appended at the end of the main body of the report. Most readers will not wish to study these in detail, at least during their first reading of the report. The use of appendices is especially valuable when two classes of reader are being served. The main body of the report can be aimed at a general reader who is mainly interested in the qualitative conclusions, while the appendices can satisfy someone doing similar work and having a more detailed interest in the methods used.

¶ 13. *References.* Any published works to which reference has been made should be listed here in the order in which they appear in the body of the report. The standard form for each reference is: 'Author's name and initials, title of paper, title of journal or book, publisher, volume number, page number, year of publication.' The list of references normally forms the final appendix.

The length of each section can vary from zero to many pages depending on how much work is being reported. The aim should be to include only that information which a reader will require.

Students have to be especially careful to visualise their readers' reactions. They will often, to save work, be asked to omit certain sections entirely. Frequently no description of the apparatus or method of operation is required. Such incomplete reports allow the teacher to check quickly whether a standard experiment has been completed satisfactorily, since he is already familiar with the apparatus.

Building up the report

The first draft of a report cannot be written in a single operation. It is helpful to build it up gradually in the following stages

(a) Jot down the reasons for doing the work and for writing the report
(b) Sketch out a list of items to be included in each chapter or section. The summary, introduction and conclusions can be left until later.
(c) Arrange these items in a logical order, transferring some to different sections if necessary.
(d) Make sure that sufficient information is available. The need for more calculation or experiment might be seen at this point.
(e) Check that the report as outlined will fulfil the original objectives
(f) Expand each list of items into a section, maintaining a balance of length between the different parts.
(g) Draw the graphs and other illustrations in the form in which they are to appear in the finished report. Number them and choose titles.
(h) Write out the summary, introduction and conclusions.
     A first draft of the report will now be ready for critical examination.

Style of writing

It is normal to write reports in the third person past tense, e.g. 'the diameter was measured' rather than 'I measured the diameter'. This practice results in impersonal, modest reports and avoids difficulties when reporting the combined operations of a group.

The use of simple words in simple sentences will make a report pleasant to read. The subject matter, not the style of writing, should hold the reader's attention. Superlatives such as 'greatest', 'best', and 'most' should normally be reserved for maxima in the mathematical sense. Where possible, clichés and catch phrases should be avoided. Advice on both style and acceptable grammar can be obtained from any of a number of standard books; for example, Vallins [2] and Gowers [3].

It is often convenient to abbreviate frequently recurring phrases. Unless the abbreviation is very well known, e.g. SFC for Specific Fuel Consumption, the phrase should be written out in full the first time it occurs and the abbreviation given in brackets. For example, early in a report we might write 'The work undertaken in the Pressure Test Facility (PTF) was particularly successful'. Then for the remainder of the report use the abbreviation 'PTF' without further explanation.

Single words should not normally be abbreviated—it is not acceptable to write 'gas turb.' instead of 'gas turbine' or 'lab.' for 'laboratory'. Common exceptions to this rule are 'Fig.', 'Eq.' and 'Ref.' for 'Figure', 'Equation' and 'Reference' respectively.

Colloquial abbreviations such as 'can't' and 'shouldn't' are not acceptable in a technical report.

Whole numbers up to ten, or perhaps twelve, should be written as words, above that as numbers, e.g. 'seven' but '17', though it is preferable that presentation of numbers should not be 'mixed' i.e. '7 up to 13' not 'seven up to 13'.

A single series of numbers, i.e. Figure 1, Figure 2, etc., should be used to refer to all the illustrations. Some authors use a different series for each type of illustration such as photographs, line drawings and graphs, but this method of labelling has no real advantage and makes the report discontinuous and less attractive.

It is most unlikely that the first draft of a report will be satisfactory. It will be necessary to go carefully through it several time to ensure, first, a good balance between the length allotted to different items and secondly, that each sentence will convey the desired meaning to the person for whom it is written. Some words such as 'heat' and 'strain' have different meanings for engineers and others. Each word has to be selected carefully to perform a particular task. No two words or phrases carry exactly the same meaning. 'A half' is not the same as '0·5', 'rapidly' is not interchangeable with 'quickly'.

Some groups of words can have a particular meaning which might not be the one intended. A simple experiment will show the difference between 'When I look into your eyes, time seems to stand still', and 'You have a face that would stop a clock'. 'High speed engine test' can have two meanings. If necessary any doubtful or ambiguous sentences should be re-phrased.

Further reading of the report might reveal excessive repetition of a single word or phrase. For example, a number of sentences might be found to start with 'Next this was done ...'. If possible these irritants should be changed, but not at the expense of clarity.

One last necessity is a check on spelling. It must be realised that poor spelling will give an impression of slovenliness, not genius, and will reduce confidence in the report. Where the report is to be typed, it is not the typist's duty to correct grammar and spelling; the total responsibility lies with the author.

Illustrations

A good drawing can transmit engineering information very economically. Before producing a drawing for a report the author must decide what information it should convey and what ability his reader is likely to have in interpreting drawings. Detail and assembly drawings as used for manufacture are not normally suitable for technical reports. For instance, in reading about some tests on the temperature distribution in a diesel engine piston, the reader should not be distracted by dimensions and tolerances of every feature of the piston. He should be given a general impression of the shape and size and details of the temperature measuring points.

In an effort to help the reader, pictorial views are often used. These are difficult to produce and are not very efficient except for showing the relative positions of features of apparatus—information which is frequently irrelevant.

More useful are semi-pictorial diagrams; electronic circuit diagrams are the best-known example but similar techniques work well in other fields. For example, flow diagrams can illustrate chemical processes or computer programs.

Whatever type of drawing is chosen, it should be produced in boldly ruled ink-lines.

Suppose a performance test has been made on a small oil fired boiler. An attempt to illustrate the equipment might result in something like Fig. 7.1. Here

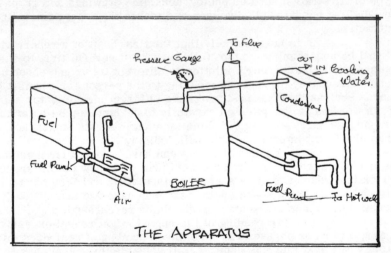

FIG. 7.1   An illustration which conveys little

an unskilled attempt at an isometric view has been made giving an idea of the general appearance of the plant. A broader nib should have been used and the features should have been labelled in block letters.

No reliance can be put on the relative sizes and positions of the parts as these might well have been distorted to fit everything in the diagram.

Figure 7.2 represents the same boiler system. This diagram was much easier to draw. The principle of operation is now obvious and the quantities measured can be seen easily. Attention has been paid to the line thickness and the lettering. Finally, a meaningful title has been used.

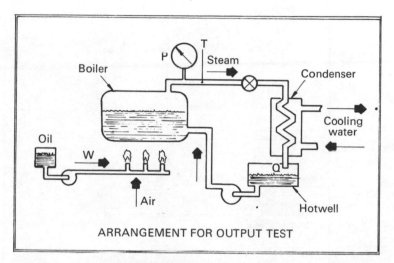

FIG. 7.2   A more helpful diagram

Graphs

A reader will wish to obtain qualitative rather than quantitative information from a graph so, in printed reports, only enough reference lines to establish the scale are included. The curves should be drawn in thick lines to enable them to stand out against the background. The general shape of the curves can then be seen at a glance. (Graphs used during calculations and in data sheets have different purposes and are drawn with fine lines on a detailed reference network.) Where a graph is used to enable calculated and measured values to be compared, it is normal to make the curve represent the calculated values, and to show the measurements as individual points. A variety of symbols, e.g. crosses, squares and triangles, can be used to distinguish different sets of results.

Where a number of graphs of similar results are included in a report, the same scale should be used for each to simplify comparisons. The scales should, in any case, be chosen to make each division of the reference grid equal to one, two, five or ten units of the variable rather than the less intelligible three, four, seven, etc., units. If a graph is drawn with a 'false origin' a note should draw the reader's attention to it to avoid risk of misunderstanding. More detailed advice on drawing graphs is given in a booklet [4] issued by the American Society of Mechanical Engineers.

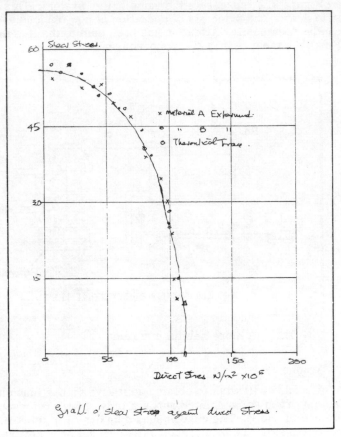

FIG. 7.3   A badly drawn graph

Figure 7.3 is a badly drawn graph which attempts to show how the results of a particular experiment on yielding under compound stresses compare with a theoretical prediction.

The scales are badly chosen—space is wasted on the horizontal axis and the divisions are awkward. It is difficult to distinguish the theoretical from the experimental values. Both the curve and the plotted points are far too feeble—they do not stand out clearly against the reference grid. Worst of all, the curve has been drawn freehand instead of with the help of French curves. The units of the horizontal axis 'stress $N/m^2 \times 10^6$' are ambiguous—it could be read as either (stress $\times 10^6$) N/m^2 or stress in ($10^6 \times N/m^2$) i.e. in MN/m^2.

     A better attempt at displaying the same data is shown in Fig. 7.4. As well as avoiding the faults listed above a more helpful title has been added and attention has been paid to the lettering.

     When submitting a graph drawn on graph paper, particular care is necessary to make the curves and points stand out. A skeleton grid should be drawn in black to contrast with the printed network. The final effect will appear similar to Fig. 7.4 but with the printed grid forming a coloured background.

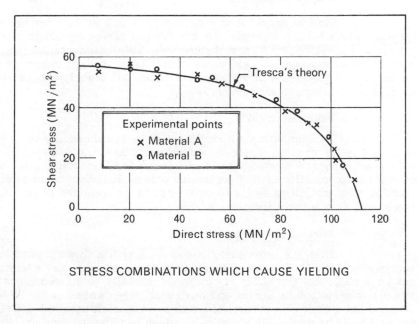

FIG. 7.4   A better presentation of the same data

Shorter communications

     Many situations do not call for a full report as outlined above. Details of work should only be included when it is certain that they will assist the reader. Frequently it is known that the reader will only require, for example, details of a particular set of apparatus or possibly the results of a certain series of measurements. In such cases the author should arrange the required material in a logical order and refine it as though it were part of a full report. He must be careful to include sufficient related details to enable his reader to make intelligent use of the main information. For example, when giving performance figures for a motor car engine it should be stated whether or not a silencer, dynamo, cooling fan, etc., are fitted.

The most common form of written communication is the letter. Here time does not normally allow rough drafts and layouts to be made. The relevant information has to be collected and assembled mentally before writing starts. Experience is necessary to master the technique. At first every letter should be re-read to think of improvements that could be made if it were to be re-written. The next letter can then incorporate those improvements.

7.4   Speaking

Skill in speaking is required daily and not just for the occasional formal lecture but most people are unwilling to accept criticism of their effectiveness at speaking, even from themselves. This is unfortunate, because performance can be improved considerably when a little thought is given to the techniques. To be able to put a point of view or explain an idea clearly and succinctly is a considerable advantage in all careers.

Informal speaking

In discussions, either direct or by telephone, the technique is similar to that of letter writing. The relevant information must be assembled mentally before speaking and no rough drafts are possible. Fortunately, the listener can easily ask for clarification of particular points. It is important to think *before* starting to speak. 'It is better to be silent and let someone think you a fool than to speak and let him know for certain.'

Formal speaking

Even the most experienced speaker has to work hard to prepare a successful lecture. The preliminary work is the same as for a written report. It has been found that the effect of reading a carefully composed report word for word is usually disastrously boring and unnatural. Even worse is the effect of reciting a memorised speech. So, instead of a finished report, a detailed list of headings and topics should be made. This outline can then be filled in spontaneously during the lecture.

A verbal presentation has the considerable advantage of allowing the speaker to see immediately the effect he is having on his audience. This enables him to change his approach (or, in an extreme case, his career).

Verbal communication is much slower than written and care is necessary to avoid trying to include too much information. For example, a 'quality' Sunday newspaper which occupies most readers for a couple of hours would require about forty hours to transmit by television. It follows that most of the detail given in a written report has to be omitted from a lecture. A speaking rate of about 100 words per minute should be assumed when planning a talk to last a certain time.

Because the audience cannot stop and refer back to a previous item, it is necessary to repeat the main topics. It is said that army instructors memorise the slogan 'First you tells 'em what you're going to tell 'em. Then you tells 'em. Then you tells 'em what you've told 'em.' This approach works well in practice.

Fortunately, because the audience hears a talk once only, slight errors in style which would be irritating to a reader are easily overlooked.

Manner of speaking

The subject matter must be completely understood before starting in order to give confidence. However, there is no need to memorise details of the order in which each topic will be made; a set of notes should be used quite openly. A sensible student will practice formal speaking fairly early in his career when it will be found that the audience is tolerant and sympathetic. To help and encourage him the Professional Institutions run junior sections. An ambitious engineer should take the trouble to tape record a talk and then play it back several times to himself. If he recovers his self-confidence he can then force himself to avoid excessive repetition of his favourite phrases or mannerisms and to speak more slowly and clearly at his next attempt. Even if his own style does not improve, he might be more sympathetic to other speakers.

If this chapter has made successful communication appear to be hard work, then it has conveyed the correct impression. There are no short cuts but the effort to succeed is well worth making. Many people make a good living communicating on behalf of others. The young engineer should learn the art in order to make the most advantageous use of his traditional training throughout his career.

LIST OF REFERENCES

1. BS 1991: PART 1: 1967, *Letter Symbols, Signs and Abbreviations,* 1967.
2. G. H. VALLINS, *Good English,* Pan Books, 1968.
3. SIR ERNEST GOWERS, Plain Words, HMSO, 1973.
4. *An A.S.M.E. Paper, American Society of Mechanical Engineers, Manual MS-4,* April 1964.

Appendix 1

1. The Principle of Dimensions

An equation modelling a real event must be independent in form from the system of dimensions used to measure all the variables. An example is provided by the Bernoulli equation for a flow of fluid which is

$$p/\rho + gz + \frac{1}{2}V^2 = K \qquad \text{(A.1.1)}$$

Using the M, L, T system of denoting the dimensions of mass, length and time respectively, then the dimensions of gz are:

$$LT^{-2}L = L^2T^{-2}:$$

the dimensions of $\frac{1}{2}V^2$ are:

$$(LT^{-1})^2 = L^2T^{-2}:$$

and the dimensions of p/ρ are:

$$\frac{(MLT^{-2}L^{-2})}{ML^{-3}} = (L^2T^{-2})$$

The dimensions of all four terms of the Bernoulli equation are seen to be identical and, further, K must have the same dimensions. Hence a change in the system of units changes each term by an identical factor whilst the form of the equation is unchanged.

Equation (A.1.1) can be put in a non-dimensional form. Just one arrangement is:

$$\frac{p}{\rho K} + \frac{gz}{K} + \frac{1}{2}\frac{V^2}{K} = 1 \qquad \text{(A.1.2)}$$

each one of the three terms on the left-hand side being dimensionless, as is the right-hand side.

2.   Buckingham's Π Theorem

The most general form of equation (A.1.1) is

$$f_1(p, \rho, z, g, V, K) = 0 \qquad\qquad (A.1.3)$$

the form of the function f_1 being unspecified. In equation (A.1.3) there are six variables and they require three dimensions to fix their numerical values, these dimensions being M, L and T.

The Π theorem of Buckingham says that these six variables can be rearranged into three non-dimensional variables, Π_1, Π_2, and Π_3, so that,

$$f_2(\Pi_1, \Pi_2, \Pi_3) = 0$$

and where the Π's are products of the variables. A general rule, to be applied with precaution (Chap. 3[1]), is that the number of non-dimensional groups is equal to the number of variables less the number of dimensions. In this example the number of variables is six and the number of dimensions three, so that the three groups needed can be the three terms in equation (A.1.2).

This is a bare statement of the Π-theorem. There is much detail in its full application; such detail can be obtained from, for example, (Chap. 3[1]).

Appendix 2 Outline of error analysis

A. 2. 1   Introduction

As it was stated in Chapter 5, the purpose of this Appendix is to describe some elementary techniques of evaluating experimental observations affected by errors and other random factors. It is not intended here to embark on an extensive review of statistical methods, but merely to outline, and to demonstrate the use of a few of the most essential procedures and formulae furnishing them with some fundamental definitions and statements. Several illustrative examples are interjected individually in places with the view of showing how the immediately preceding material can be applied in real situations. One entire section, namely A. 2. 7, constitutes nothing else than just one single exercise covering the material presented in all previous sections. This should be instrumental as processing of a more or less complete experimental case is considered there.

Imagine a sequence of measurements or (speaking in more general terms) observations whose nominal values are expected to remain constant. In fact the observations, as pointed out in Chapter 5, will most likely exhibit a random experimental scatter. For example, the actual numerical values of a mechanical property of material (e.g. yield stress) may vary from specimen to specimen, even though the specimens are supposed to be technically identical, i.e. made from the same material and to the same manufacturing specifications. Similarly, a series of identical automobile engines when tested on a testing rig are likely to produce somewhat different performance parameters (e.g. fuel consumption, maximum power, torque etc.) for the same test conditions. Also lives of individual electronic transistors, from the same production batch, will not be exactly equal. In all these cases the actual items whose performance is being monitored may be slightly different and there may be experimental errors in the relevant observations. Both components are inherently random, i.e. without a coherent pattern. However, in spite of that, a rational quantification of such randomly scattered observations is possible thanks to several simple statistical rules which will be discussed below.

A. 2. 2   Presentation of observations

Any set of randomly scattered observations, say measurements affected by errors, can be presented as a *histogram* provided a sufficiently large number of observations has been recorded.

Denote in general the observations by X_k and define intervals ΔX_k centred around X_k where subscript $k = 1, 2, \ldots, N$ with N = total number of intervals. Now count the number of observations that fall into every interval ΔX_k and call it $m_k = frequency\ of\ occurrence$ per interval. Plotting m_k for all intervals ΔX_k across the entire range of observations X_k and presenting the diagram in the form of a collection of vertical bars, as shown in Fig. A. 2. 1, constitutes the histo-

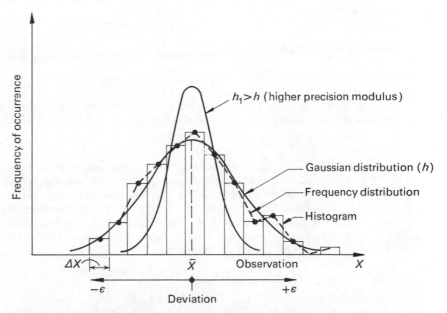

FIG. A. 2. 1   Frequency distribution (histogram)

gram. A line plot joining consecutively all top centres of the histogram bars is known as the *frequency distribution* plot which can also be seen in Fig. A. 2. 1 The latter plot, as will be shown in section A. 2. 4, is relevant when experimental frequency distributions are compared with theoretical (idealised) models.

A. 2. 3   The mean and the measures of deviation

One natural tendency when appraising scattered data is to calculate first the respective *arithmetic mean,* hereafter called simply the mean. Incidentally, it can be shown by rigorous statistical speculations to be the *most probable* value, so the mean appears to represent a very important statistical parameter.

Before the well known formula for the mean is once again written out here, one should be able to distinguish two possible types of 'collections of items or events' for which the mean may be required. Firstly, one may be faced with an almost infinitely large *population* whose size has but negligible effect upon the mean. Such a mean will be denoted by μ. It is a sort of absolute mean. Secondly, there is a sample mean

$$\overline{X} = \frac{X_1 + X_2 + \cdots + X_n}{n} = \frac{1}{n} \sum_{1}^{n} X_k \tag{A.2.1}$$

which refers to a finite set of n observations $X_1, X_2, \ldots, X_n$. The size n does affect $\overline{X}$. Obviously for very large samples $\overline{X} \to \mu$.

Using the concept of frequency of occurrence,

$$\overline{X} = \frac{m_1 X_1 + m_2 X_2 + \cdots + m_N X_N}{m_1 + m_2 + \cdots m_N}, \tag{A2.1a}$$

where $n = m_1 + m_2 + \cdots + m_N$.

In order to quantify a measure of deviation from the mean for the same sample of n observation one wishes to define the basic deviation

$$\epsilon_k = X_k - \overline{X}. \tag{A2.2}$$

It would be futile to try to calculate the arithmetic mean of the basic deviations as it can be shown on sight that $\overline{\epsilon}$ equals identically zero. In electronics one frequently meets with alternating signals and it was for this purpose that the concept of the *root-mean-square* (RMS) value was invented. To obtain the RMS deviation, one squares all deviations ϵ_k, sums them together, divides the resultant sum by the total number of observations, and finally takes the square root the result in order to return to the correct order of magnitude.

$$D = \left(\frac{\epsilon_1^2 + \epsilon_2^2 + \cdots + \epsilon_n^2}{n} \right)^{1/2} = \left[\frac{1}{n} \sum_{1}^{n} \epsilon_k^2 \right]^{1/2} \tag{A.2.3}$$

The quantity D is called the *RMS deviation* of a sample.

Statisticians argue that it is not strictly correct to use the total number of observations n for the purpose of finding the measure of deviation for a sample of observations possesses n *degrees of freedom* one of which is removed by calculating $\overline{X}$. One should, they say, average the sum of squared deviations over the number of remaining degrees of freedom, i.e. by $n - 1$. Therefore the accepted measure of deviation of a finite sample is:

$$S = \left[\frac{1}{n-1} \sum_{1}^{n} \epsilon_k^2 \right]^{1/2} = \left(\frac{n}{n-1} \right)^{1/2} D, \tag{A.2.4a}$$

which is referred to as the sample *standard deviation* and its square as the sample *variance*. The factor $n/(n-1)$ is called *Bessel's correction*.

A similar quantity evaluated for a population would be the population standard deviation σ representing the measure of deviation from μ.

Both sample mean $\overline{X}$ and sample standard deviation S vary from sample to sample, but they tend to μ and σ for $n \to \infty$. Also the difference between the RMS deviation D and S becomes very small for large samples: $(S - D)/S = 0\cdot025$ for $n = 20$ and it tends rapidly to zero with $n \to \infty$.

There is an alternative measure of deviations from the mean, namely the *mean absolute deviation* (MAD)

$$|\epsilon| = \frac{1}{n} \sum_{1}^{n} |\epsilon_k|, \tag{A.2.4a}$$

which for obvious reasons is never zero save for the trivial case of all $\epsilon_k = 0$.

One interesting feature of the mean $\overline{X}$ is that the sum of squared deviation $\sum \epsilon_k^2$ can be shown to be always minimum (see also section A.2.10) which is another indication that $\overline{X}$ must indeed be the most probable value.

A.2.4 The Gaussian law of errors as a theoretical model for assessing observations

Seemingly it is a law of nature that almost all unbiased observations assume their frequency distribution plots similar to that portrayed previously in Fig. A.2.1. There is a well-known analytical curve, known as the Gaussian curve, whose symmetric bell-shaped appearance renders it particularly suitable as a 'templet' for judging the experimental frequency distributions. Changing in Fig. A.2.1 the abscissa coordinates from X to ϵ such that the origin of ϵ's coincides with $\overline{X}$, the frequency of occurrence (signified by y) becomes a function of ϵ.

Apply now the Gaussian curve to a population in order to get the desired theoretical model for the frequency distribution of deviations $\epsilon = X - \mu$,

$$y(\epsilon) = \frac{h}{\sqrt{\pi}} e^{-h^2 \epsilon^2}. \tag{A.2.5}$$

Symbol h is here a parameter representing the geometric 'spread' of the Gaussian curve. A formal derivation of Equation (A.2.5), as a frequency distribution model, is possible using the conventional probability argument. It can be found, for example, in Beers (Ch.5[2])

There is obviously an inconsistency in applying a continuous function (Equation (A.2.5)) as a model to a discrete collection of observations. The inconsistency is, however, of secondary importance as long as the number of observations is sufficiently large.

Adopting the Gaussian model without further reservations, one specifies now a deviation interval $\Delta\epsilon$ centred around some deviation ϵ. The ratio of the number of deviations that can be found within $\Delta\epsilon$ to the total number of deviations involved n,

$$P(\epsilon) = \frac{m}{n}$$

is the *probability* that an observation X will differ from its mean $\overline{X}$ or μ (whichever case it may be) by $\epsilon \pm \Delta\epsilon/2$. The probability $P(\epsilon)$ can be hence identified with the non-dimensionalised frequency distribution $y(\epsilon)$ so that the number of observation whose deviations fall into the interval $\Delta\epsilon$ is:

$$ny(\epsilon)\Delta\epsilon = \frac{nh}{\sqrt{\pi}} e^{-h^2\epsilon^2} \Delta\epsilon$$

The population standard deviation therefore reads:

$$\sigma = \lim_{n\to\infty} \frac{1}{n-1} \int_{-\infty}^{+\infty} \epsilon^2 \frac{nh}{\sqrt{\pi}} e^{-h^2\epsilon^2} d\epsilon = \frac{1}{h\sqrt{2}}. \tag{A.2.6}$$

The parameter h is responsible, as already mentioned, for the shape of the frequency distribution curve. The higher its value the more peaky the distribution curve which in turn means a more precise set of observations. A high frequency of small deviations indicates a high proportion of data lying close to the mean. Hence the parameter h bears the name *precision modulus*.

Because of its direct application in the theory of errors the Gaussian model appears in the literature under the name of *Gaussian law of errors*. The function $y(\epsilon)$ is correspondingly called the *Gaussian or normal frequency distribution*.

Other forms of the frequency distributions may emerge sometimes in practice especially when a set of observations is for some reasons biased in one or more directions. This is why there is a need for different distribution models to cope with such non-Gaussian situations. Here one should mention the bionomial, hypergeometric, and Poisson distributions. Unlike the normal distribution, these distributions are discrete since they entail various forms of series expansions. Their applications range from the sampling procedures through problems of failure to establishing the probability of accidents in time. The reader is at this point referred to literature, for example Chatfield [1] or Guttman et al. [2], for further details. Statistical *significance* tests and tests for *goodness-of-fit* are based on two important continuous distribution models: (i) Student's t-distribution and (ii) the χ^2-distribution, respectively. The significance tests receive some futher attention later in section A.2.11 but regarding the goodness-of-fit procedures the reader once again is advised to consult literature for particulars (e.g. [1] and [2]).

A.2.5   The probable deviations and rejection of observations

Turning now to practical situations, one can think of a situation where two limits, say a_1 and a_2, are specified so that all 'acceptable' deviations must be contained within $a_1 \leqslant \epsilon \leqslant a_2$. The present desideratum is the probability of occurrence of such deviations. For a population the probability equals the area under the Gaussian curve between the two limits a_1 and a_2. When a_1 and a_2 become $-\infty$ and $+\infty$, respectively, then the area is *unity* which means 100 per cent probability since all deviations are acceptable.

A customary non-dimensional deviation

$$Z = \frac{X - \mu}{\sigma} = \frac{\epsilon}{\sigma} \qquad (A.\,2.\,7)$$

transforms Equation (A. 2. 5) into

$$y(Z) = \sigma y(\epsilon) = \frac{1}{\sqrt{2\pi}}e^{-z^2/2}, \qquad (A.\,2.\,8)$$

where h has been substituted by Equation (A. 2. 6).
The probability area under $y(Z)$ is

$$A = \frac{1}{\sqrt{2\pi}} \int_{\alpha_1}^{\alpha_2} e^{-z^2/2}dZ, \qquad (A.\,2.\,9)$$

where $\alpha_1 = a_1/\sigma$ and $\alpha_2 = a_2/\sigma$ are the non-dimensional limits of integration. As is well known, there is no explicit solution to integral (A. 2. 9) and its values are available in a tabulated form. An abridged version of the tabulated values is given here as Table A. 2. 1 where $\frac{1}{2}A$ stands for integral (A. 2. 9) evaluated between the limits $\alpha_1 = 0$ and $\alpha_2 = \alpha$. This is naturally sufficient as the Gaussian curve is *symmetric* about the origin.

TABLE A. 2. 1*

α	$\frac{1}{2}A$	α	$\frac{1}{2}A$	α	$\frac{1}{2}A$
0·00	0·0000	0·50	0·191 5	1·60	0·445 2
0·05	0·019 9	0·60	0·225 7	1·80	0·464 1
0·10	0·039 8	0·70	0·258 0	2·00	0·477 2
0·15	0·059 6	0·80	0·288 1	2·20	0·486 1
0·20	0·079 3	0·90	0·315 9	2·60	0·495 3
0·25	0·098 7	1·00	0·341 3	3·00	0·498 7
0·30	0·117 9	1·10	0·364 3	3·40	0·498 8
0·35	0·136 8	1·20	0·384 9	4·00	0·500 0
0·40	0·155 4	1·30	0·403 2		
0·45	0·173 7	1·40	0·419 2		

* Values extracted (by permission) from: Pearson and Hartley *Biometrika Tables for Statisticians*, Camb. Univ. Press, 1966.

Using Table A. 2. 1 it is evident that symmetric deviation limits $\alpha = \pm 1$ imply, for a population with normal distribution, that 68·26 per cent of all observations will be acceptable. Conversely, if probability of 68·26 per cent is required for observations to occur within symmetric limits centred around the origin $\epsilon = 0$ then it becomes necessary to reject all those observations whose deviations from mean μ fall outside the range $-\sigma \leqslant \epsilon \leqslant \sigma$. Similarly, the probability of 95·44 per cent imposes limits $-2\sigma \leqslant \epsilon \leqslant 2\sigma$.

A rapid *rejection* criterion can be evolved from that. The stringency of such a criterion depends solely on the probability desired which itself depends on the individual experimental circumstances. The limits on deviations for the rejection criterion need not be symmetric, of course.

Example (a)

A certain dimension of a mass produced mechanical part has a nominal value of 100 mm with acceptable tolerances of ± 1 mm. Several large samples have been examined resulting in the population mean $\mu = 100\cdot3$ mm and the population standard deviation $\sigma = 0\cdot5$ mm. Calculate the percentage of production that is likely to be rejected.

Limits:

$$a_1 = \frac{(100 - 1) - 100\cdot3}{0\cdot5} = -2\cdot6$$

$$a_2 = \frac{(100 + 1) - 100\cdot3}{0\cdot5} = 1\cdot4$$

Using Table A. 2.1, and recalling the symmetry of the normal distribution, the rejected percentage is

$$100 \left[1 - (0\cdot419\,2 + 0\cdot495\,3)\right] = 8\cdot6 \text{ per cent}$$

A set of observations may sometimes contain an odd value which seems inconsistent with the rest of the data. Often there is a temptation to discard the suspect result. How does one decide whether or not to reject it? This ma be an important question especially for small samples where the probability of occurrence of large deviations is small and retention of such incoherent observations may distort the overall result or conclusion. The answer to this problem is not always easy as clear cut reasons for taking the correct decisions are rarely at hand.

First of all, an odd looking observation may be in fact quite legitimate and it is always prudent to check this point. Secondly, a decision is needed as to what probability of occurrence is acceptable or desirable. The choice is rather individual, but unless there are other special reasons one may be justified in rejecting those observations whose deviations are less than 50% probable to happen. This particular rule is fairly popular amongst experimental physicists; Example (b) demonstrates its application.

Example (b)

An experimental sample consisting of 13 observations of the relative humidity index in some stable environment resulted in the following readings: 49, 51, 50, 47, 52, 57, 53, 48, 50, 52, 54, 51 and 49. At first sight the readings 47 and 57 are potentially suspect. Assuming that no special reasons exist for the deviations, apply the 50 per cent probability criterion to see if any of the two readings could be left out.

Mean:

$$\overline{X} = \frac{663}{13} = 51$$

Standard deviation:

$$S = \sqrt{86/12} = 2 \cdot 68$$

50 per cent probability criterion:

$$n(1 - A) = \frac{1}{2}$$

$$A = \frac{2n - 1}{2n} = \frac{25}{26} = 0 \cdot 961\,5$$

for which value interpolation of Table A.2.1 gives $\alpha = \pm 2 \cdot 07$. Hence the maximum acceptable deviations are

$$\epsilon_{max} = \pm 2 \cdot 07 \times 2 \cdot 68 = \pm 5 \cdot 55.$$

The reading 57, whose deviation from mean is $57 - 51 = 6 > 5 \cdot 61$, can be rejected but the reading 47 giving deviation from mean $47 - 51 = -4 > -5 \cdot 55$ should be retained.

A.2.6   Checking the applicability of the Gaussian model

Before applying the Gaussian model to the experimental distribution under investigation one must first establish that there is a satisfactory resemblance between the two distributions. Some experimental distributions can be inherently biased and the Gaussian model would be irrelevant. They may be skewed, excessively peaky (leptokurtic) or excessively flat (platykurtic) as presented in Fig. A.2.2. Apart from superimposing the distribution and the model on one another graphically and exercising a rather subjective judgement about their mutual resemblance, one can perform a simple statistical test based on two typical 'shape' factors: *the skewness factor* λ and the *flatness factor* κ which for the normal distribution are known to be 0 and 3, respectively. Both λ and κ are related to the RMS deviation D (cf. Eq. (A.2.3))

$$\lambda = \frac{1}{nD^3} \sum_1^n \epsilon_k^3; \quad \kappa = \frac{1}{nD^4} \sum_1^n \epsilon_k^4 \qquad \text{(A.2.10)}$$

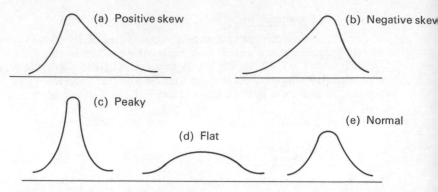

FIG. A. 2. 2   Abnormal distributions

It is probably in order to accept as near enough normal all those experimental distribution for which $-0 \cdot 3 < \lambda < 0 \cdot 3$ and $2 \cdot 5 < \kappa < 3 \cdot 5$. Different distribution models should be sought, as suggested in Section A. 2. 4, if the above limits are exceeded.

A goodness-of-fit test was also mentioned in section A. 2. 4. The essence of this test rests on evaluating the parameter

$$\chi^2 = \sum_1^n \frac{(\text{observations} - \text{expected values})^2}{\text{expected values}}$$

where the expected values could be, for example, the postulated Gaussian model. The χ^2-test should be regarded as an alternative or complementary test for checking the applicability of the Gaussian model. The computed χ^2 is normally compared with its value resulting from the χ^2 distribution for an appropriate probability and corresponding number of degrees of freedom (depending on the size of the sample). For working details of the χ^2-test consult the recommended literature.

A. 2. 7   An exercise involving material presented in sections A. 2. 2–2. 6

It may be useful to illustrate the application of the material presented thus far on a practical example.

Suppose that it was required to conduct a hydraulic experiment during which a certain pressure should have been maintained constant and equal to 10 bar ($1 \text{ bar} = 10^5 \text{ N/m}^2$). However, this condition proved impossible to secure and, as seen in Table A. 2. 2, where results of $n = 50$ consecutive readings are tabulated, some appreciable pressure fluctuations were observed. The experimenter feels that in his report he must document this particular source of error as it affects his further results. He therefore carried out the following calculations:

1. *Data presentation:* See Table A. 2. 2 where m = number of occurrences per interval $\Delta X = 0 \cdot 1$ bar; Fig. A. 2. 3 contains the respective histogram.

TABLE A.2.2

X	m	Σm	mX	$\epsilon = \bar{X} - X$	ϵ^2	ϵ^3	ϵ^4	$m|\epsilon|$	$m\epsilon^2$	$m\epsilon^3$	$m\epsilon^4$
9·6	1	1	9·6	−0·45	0·2025	−0·0911	0·0410	0·45	0·2025	−0·0911	0·0410
9·7	2	3	19·4	−0·35	0·1225	−0·0429	0·0150	0·70	0·2450	−0·0858	0·0300
9·8	6	9	58·8	−0·25	0·0625	−0·0156	0·0039	1·50	0·3750	−0·0936	0·0234
9·9	7	16	69·3	−0·15	0·0225	−0·0034	0·0005	1·05	0·1575	−0·0238	0·0035
10·0	10	26	100·0	−0·05	0·0025	−0·0001	0·0000	0·50	0·0250	−0·0010	0·0000
10·1	8	34	80·8	0·05	0·0025	0·0001	0·0000	0·40	0·0200	0·0008	0·0000
10·2	8	42	81·6	0·15	0·0225	0·0034	0·0005	1·20	0·1800	0·0272	0·0040
10·3	4	46	41·2	0·25	0·0625	0·0156	0·0039	1·00	0·2500	0·0624	0·0156
10·4	3	49	31·2	0·35	0·1225	0·0429	0·0150	1·05	0·3675	0·1287	0·0450
10·5	0	49	0	0·45	0·2025	0·0911	0·0410	0	0	0	0
10·6	1	50	10·6	0·55	0·3025	0·1664	0·0915	0·55	0·3025	0·1664	0·0915
Σ	50	—	502·5	—	—	—	—	8·40	2·1250	0·0902	0·2540

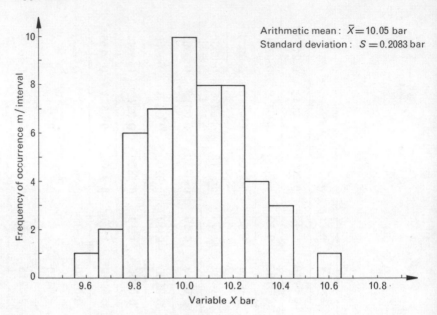

FIG A. 2. 3   Histogram

2. *Mean:* (Eq. (A. 2. 1a)): $\overline{X} = 502 \cdot 5/50 - 10 \cdot 05$ bar.
3. *RMS deviation:* (Eq. (A. 2. 3)): $D = \sqrt{2 \cdot 125/50} = 0 \cdot 206\ 2$ bar.
4. *MAD:* (Eq. (A. 2. 4)): $|\epsilon| = 8 \cdot 40/50 = 0 \cdot 168$ bar.
5. *Standard deviation:* (Eq. (A. 2. 4a)): $\sigma = \sqrt{50/49}\ 0 \cdot 206\ 2 = 0 \cdot 208\ 3$ bar.
6. *Precision Modulus:* (Eq. (A. 2. 6)): $h = 1/0 \cdot 208\ 3\ \sqrt{2} = 3 \cdot 30$ bar^{-1}.
7. *Gaussian curve:* Probability per interval $\Delta\epsilon = 0 \cdot 1$

$$y \Delta\epsilon = \frac{3 \cdot 39 \times 0 \cdot 1}{\sqrt{\pi}} e^{-3 \cdot 39^2 \epsilon^2} = 0 \cdot 191\ e^{-11 \cdot 5 \epsilon^2}$$

See Fig. A. 2. 4.

8. *Skewness and flatness factors:* (Eq. (A. 2. 10):

$$\lambda = \frac{0 \cdot 0902}{50 \times 0 \cdot 2062^3} = 0 \cdot 206; \quad \kappa = \frac{0 \cdot 2540}{50 \times 0 \cdot 2062^4} = 2 \cdot 81$$

Neither λ nor κ are ideally Gaussian (0 and 3, respectively) but within the limits recommended in section A. 2. 6.

A. 2. 8   The standard error

If random samples, each of size n, are drawn from a population whose

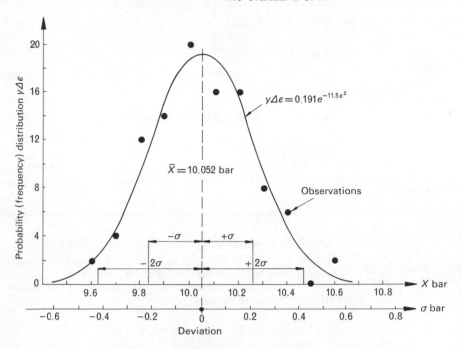

FIG A. 2. 4   Gaussian model fit

mean is μ and standard deviation is σ, then the sample means $\overline{X}_i = \mu$ and their standard deviation, known as the *standard error*, is

$$\sigma_m \cong \frac{\sigma}{\sqrt{n}}. \tag{A. 2. 11}$$

Further, if similar random samples are drawn from a population with a normal distribution, then the distribution of means $\overline{X}_i$, is also normal with an identical standard error σ_m.

Equation (A. 2. 11) is only approximate but its accuracy improves with the sample size n. It is considered to be correct for $n > 20$.

Example (c)

In order to establish a routine checking of the concentration of a certain chemical in a substance manufactured on a production line, one takes a series of measurements at regular intervals. Knowing from previous samples that there is no systematic error involved and that the standard deviation per sample is 2·5 per cent, calculate what should be the minimum size of the regular test samples so that the standard error of the concentration be kept under 0·5 per cent?

Using Eq. (A. 2. 11)

$$n \geqslant \left(\frac{\sigma}{\sigma_m}\right)^2 = \left(\frac{2 \cdot 5}{0 \cdot 5}\right)^2 = 25$$

Minimum sample size is 25.

A. 2. 9   Propagation of errors

To determine experimentally the power of a gasoline motor it is necessary to measure the torque on, and the rotational speed of, the driving shaft (cf. Appendix 1). This is a typical example where an experimental quantity is not measured directly, but calculated from a formula using several constituent measurements.

$$Q = Q(X, Y, Z, \cdots).$$

All these constituent measurements are subject to their individual errors $\Delta X, \Delta Y, \Delta Z, \cdots$ thus inflicting an error ΔQ upon the quantity Q. The objective of this section is to show how ΔQ can be estimated.

Apply in this case the well-known calculus formula for a small increment of a function of many variables

$$\Delta Q = \frac{\partial Q}{\partial X}\Delta X + \frac{\partial Q}{\partial Y}\Delta Y + \frac{\partial Q}{\partial Z}\Delta Z + \cdots, \qquad (A. 2. 12)$$

where the partial derivatives $\partial Q/\partial X$, $\partial Q/\partial Y$, $\cdots$ must be evaluated within the intervals $\Delta X, \Delta Y, \cdots$. The following Example (d) explains how the procedure works.

Example (d)

The classical simple pendulum experiment may be used to determine the gravitational acceleration g. Assuming that it is possible to measure the length of the pendulum with virtually no error at all, find what maximum error can be allowed for the time measurement in order to secure an accuracy of 1 per cent for g.

The pendulum equation is:

$$T = \pi \sqrt{\frac{L}{g}},$$

where L = pendulum length; T = period of oscillations. Hence

$$g = \pi^2 \frac{L}{T^2}$$

$$\Delta g = \pi^2 \left(\frac{\Delta L}{T^2} - 2\frac{L}{T^3}\Delta T\right)$$

$$\frac{\Delta g}{g} = \frac{\Delta L}{L} - 2\frac{\Delta T}{T} \cong -2\frac{\Delta T}{T},$$

since $\frac{\Delta L}{L} \ll \frac{\Delta T}{T}$ by assumption. In order to attain an accuracy of 1 per cent for g, the period T must be measured with an accuracy of at least $\frac{1}{2}$ per cent.

In the above Example it has been assumed that the error tolerances ΔL and ΔT are well defined. This condition could be difficult to satisfy in practice. How, for example, can one guarantee that certain error tolerances will never be surpassed? It is often safer, therefore, to replace the absolute tolerances in Eq. (A.2.12) with the respective standard error

$$\sigma_{m_Q} = \left(\frac{\partial Q}{\partial X}\sigma_{m_X}\right)^2 + \left(\frac{\partial Q}{\partial Y}\sigma_{m_Y}\right)^2 + \cdots, \tag{A.2.13}$$

which mathematical transformation can be strictly justified.

The following rules apply to the four basic algebraic operations on the constituent measurements:

1. *Addition and/or subtraction*

$$Q = a \pm b \pm c \pm \cdots$$
$$\sigma_{m_Q}{}^2 = \sigma_{m_a}{}^2 \pm \sigma_{m_b}{}^2 \pm \sigma_{m_c}{}^2 \pm \cdots$$

2. *Multiplication and/or division of powers* $(\alpha, \beta, \gamma, \cdots)$;

$$Q = a^\alpha b^\beta c^\gamma \cdots$$
$$\sigma_{m_Q}{}^2 = (a^\alpha b^\beta c^\gamma \cdots)(\frac{\alpha}{a}\sigma_{m_a}{}^2 + \frac{\beta}{b}\sigma_{m_b}{}^2 + \frac{\gamma}{c}\sigma_{m_c}{}^2 + \cdots)$$

Example (e)

Derive an expression for the standard error of the quantity Q

$$Q = A^2/\sqrt{B} = A^2 B^{-1/2}$$

in terms of the constituent σ_{m_A} and σ_{m_B}

$$\sigma_{m_Q}{}^2 = A B^{-1/2}\left(\frac{2}{A}\sigma_{m_A}{}^2 - \frac{1}{2B}\sigma_{m_B}{}^2\right) = \frac{2A}{\sqrt{B}}\sigma_{m_A}{}^2 - \frac{A^2}{2B\sqrt{B}}\sigma_{m_B}{}^2$$

A.2.10   Curve fitting by least squares

Curve fitting by eye is probably by far the most common method of fixing an emperical relationship for two sets of mutually dependent variables. For high precision experiments where the scatter of points is not too great, the procedure is in fact quite satisfactory. It may not be, however, for low precision experiments yielding widely scattered results. The process of estimating the functional dependence of an empirical relationship using a set of values for a sample is known as the *regression analysis*.

Probably the most common regression analysis is the well established *least squares* procedure where the key feature is that the sum of all deviations between the actual coordinates of the dependent variables and the coordinates of the regression curve (often called the *residuals*) *must be minimum*.

Consider a set of n experimental points, for example as in Fig. A.2.5. The respective residuals are

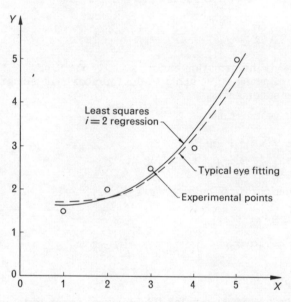

FIG A.2.5   Least squares curve fitting (example (f))

$$\delta_k = f(X_k) - Y_k \quad (k = 1, 2, \ldots, n)$$

where $(X_1, Y_1), (X_2, Y_2), \cdots, (X_n, Y_n)$ are the experimental points and $Y = f(X)$ is the regression function to be fitted. The function is a least squares regression if

$$\delta_1{}^2 + \delta_2{}^2 + \cdots + \delta_n{}^2 = \text{minimum}$$

Polynomial functions $f(X)$ are particularly suitable for this purpose

$$Y = C_0 + C_1 X + C_2 X^2 + \cdots + C_i X^i = \sum_0^i C_k X^k. \qquad (A.2.14)$$

Here i denotes the order of the polynomial. The $i + 1$ coefficients C_0 through C_i are calculated from the condition of minimum residuals,

$$R = \sum_0^i \delta_k{}^2 = \sum_0^i [f(X_k) - Y_k]^2 = \text{minimum},$$

which is satisfied by

$$\frac{\partial R}{\partial C_0} = \frac{\partial R}{\partial C_1} = \cdots = \frac{\partial R}{\partial C_i} = 0. \tag{A.2.15}$$

The necessary requirement obviously is that the number of experimental points available for the polynomial regression analysis is at least one more than the order of the polynomial to be fitted, i.e. $n \geqslant i + 1$.

Conditions (A.2.15) yield the following expressions for the coefficients of an i-th order polynomial Eq.(A.2.14).

$$C_0 n + C_1 \sum_1^n X_k + \cdots + C_i \sum_1^n X_k{}^i = \sum_1^n Y_k$$

$$C_0 \sum_1^n X_k + C_1 \sum_1^n X_k{}^2 + \cdots + C_i \sum_1^n X_k{}^{i+1} = \sum_1^n X_k Y_k \tag{A.2.16}$$

$$\cdots$$

$$C_0 \sum_1^n X_k{}^i + C_1 \sum_1^n X_k{}^{i+1} + \cdots + C_i \sum_1^n X_k{}^{2i} = \sum_1^n X_k{}^i Y_k,$$

which for a straight line $(Y = C_0 + C_1 X)$ degenerate to

$$\left.\begin{array}{l} C_0 n + C_1 \sum_1^n X_k = \sum_1^n Y_k \\[2mm] C_0 \sum_1^n X_k + C_1 \sum_1^n X_k{}^2 = \sum_1^n X_k Y_k. \end{array}\right\} \tag{A.2.17}$$

The higher the order of the polynomial the more laborious are the calculations leading to the solution of a set of $i + 1$ algebraic equations (A.2.16). There are special numerical subroutines to do the job on the computer, but it would be foolish to try always to fit the highest order polynomial possible ($i = n - 1$). A quick visual examination of the distribution of the experimental points considered, often suggests a low order polynomial (straight line, parabola, cubic) which could be quite adequate. One should realise that higher order polynomials fitted by the least squares technique frequently tend to undulate which feature may be quite wrong for the physical phenomenon that the particular regression is meant to represent.

When fitting the regression curves one must always decide early which of the interdependent variables is going to be the independent variable. The choice is not always obvious, and it seems best to call the independent variable that one which is easier to measure.

The effect of the condition of minimum residuals in least squares procedures is that the resultant Y's are statistically most probable. Figure A.2.6 shows this feature by indicating the respective frequency distributions which

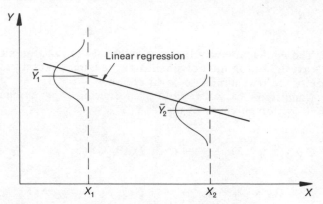

FIG A.2.6   Distribution of Y at X = const.

concept is very pertinent here. In section A.2.12 the problem of confidence intervals for the linear regression will be discussed further.

Example (f)

A geometric contour is defined by a set of approximate coordinates (X_k, Y_k), as shown in Fig. A.2.5. They are $(1, 1\cdot5)$, $(2, 2)$, $(3, 2\cdot5)$, $(4, 3)$, $(5, 5)$. Try to fit a least squares parabola to these points. The procedure is given in the table below.

Coefficients C_0, C_1, C_2							Regression			
X_k	Y_k	$X_k{}^2$	$X_k{}^3$	$X_k{}^4$	$X_k Y_k$	$X_k{}^2 Y_k$	C_0	$C_1 X$	$C_2 X^2$	Y
1	1·5	1	1	1	1·5	1·5	1·9	0·485 7	0·214 3	1·629
2	2·0	4	8	16	4·0	8·0	1·9	0·971 4	0·857 2	1·786
3	2·5	9	27	81	7·5	22·5	1·9	1·457 1	1·928 6	2·371
4	3·0	16	64	256	12·0	48·0	1·9	1·942 8	3·438 6	3·386
5	5·0	25	125	625	25·0	125·0	1·9	2·428 6	5·357 2	4·829
Σ 15	14	55	225	979	50	205				

The equations for the coefficients C_0, C_1, C_2

$$C_0 \; 5 + C_1 \; 15 + C_2 \; 55 = 14$$

$$C_0 \; 15 + C_1 \; 55 + C_2 \; 225 = 50$$

$$C_0 \; 55 + C_1 \; 225 + C_2 \; 979 = 205$$

have a solution:

$$C_0 = 1{\cdot}9, C_1 = -\,0{\cdot}485\,7;\, C_2 = 0{\cdot}214\,3$$

Hence, the least squares parabola is

$$Y = 1{\cdot}9 - 0{\cdot}4857 \; X + 0{\cdot}2143 \; X^2$$

The results are presented in Fig. A. 2. 5 where an intuitive eye fit is also super-imposed for comparison.

A. 2. 11   The significance tests

    Means of detecting whether differences between two sets of observations are due to a real reason or merely due to a chance factor are often vital in experimentation. Also two interim inspection samples from the same population are unlikely to be identical and one needs to establish whether the differences are insignificant and incidental or significant indicating progressive permanent changes in the population.

    The differences between samples are said to be significant if the probability of getting one as large as, or larger than, the observed ones is less than 5 per cent, i.e. one in twenty. The 5 per cent probability criterion is not a rigid rule, but a reasonable practical compromise. It involves the risk of a true hypothesis being discarded once in twenty samples.

    For a sample of n observations drawn from a Gaussian population the non-dimensional deviation of sample mean $\overline{X}$ from population mean μ is according to Eqs. (A. 2. 7) and (A. 2. 11)

$$Z = \sqrt{n} \; \frac{\overline{X} - \mu}{\sigma}.$$

It is not common to know the population standard deviation σ so a similar non-dimensional deviation based on the sample standard deviation,

$$t = \sqrt{n} \; \frac{\overline{X} - \mu}{S}, \qquad\qquad\qquad (A.\,2.\,18)$$

is better used.

    The frequency distribution of the deviation t is known to be for small size samples $(n < 20)$ distinctly different from the normal distribution. The cor-

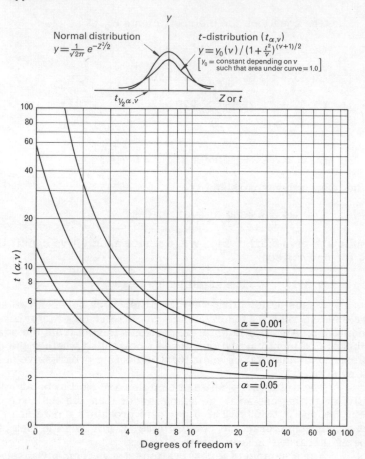

FIG A. 2. 7   t — Significance test: values for $t_{\alpha,\nu}$ [$\alpha = 0\cdot001$ (very significant); $\alpha = 0\cdot01$ (significant); $\alpha = 0\cdot05$ (rather significant)]

responding theoretical model, shown in Fig. A. 2. 7, is not as peaky as the Gaussian model and it depends strongly on the number of degrees of freedom ν of the sample concerned (cf. sec. A. 2. 3). For large ν's, i.e. large samples, the t-distribution becomes asymptotically Gaussian.

Here again the area under the t-curve represents the probability of a deviation to occur. The symbol α represents conventionally the combined areas under the t-curve from $-\infty$ to $-t_\nu$ and from t_ν to $+\infty$ (see Fig. A. 2. 7) and as such is called the *significance level* α. In line with the above probability criterion, the parameter α indicates the probability of getting the sample mean deviation $\overline{X} - \mu$ equal or greater than

$$\frac{t_{\alpha, \nu} S}{\sqrt{n}}$$

Alternatively, $1 - \alpha$ signifies the *confidence* that the sample mean deviation $\overline{X} - \mu$ will not exceed the limits

$$\pm \frac{t_{\alpha, \nu} S}{\sqrt{n}} .$$

This is in fact nothing more than an extension of the procedure evolved in section A. 2. 5 for normal distributions. Presently it is being applied to small size samples under the name of *t-significance tests*. The corresponding nomogram for three significance levels $\alpha = 0{\cdot}05, 0{\cdot}01,$ and $0{\cdot}001$ is given in Fig. A. 2. 7.

Example (g)

The lift coefficient on an aerofoil has been calculated using two independent theoretical methods A and B. For a given incidence angle, theory A predicts $C_L = 1{\cdot}81$ and theory B predicts $C_L = 1{\cdot}87$. To verify both theories a model of the aerofoil has been tested at the same incidence angle and the corresponding flow conditions (Reynolds number etc.). The following five observations have been recorded for C_L: $1{\cdot}75, 1{\cdot}80, 1{\cdot}82, 1{\cdot}72$ and $1{\cdot}76$. Apply the *t*-significance test to establish if any of the theoretical results are significantly different from the observations.

1. *Mean:* $\overline{X} = \dfrac{1{\cdot}75 + 1{\cdot}80 + 1{\cdot}82 + 1{\cdot}72 + 1{\cdot}76}{5} = \dfrac{8{\cdot}85}{5} = 1{\cdot}77$

2. *Standard deviation:* $\Sigma\epsilon^2 = 0{\cdot}0064$

$$S = \sqrt{\frac{0{\cdot}0064}{5-1}} = 0{\cdot}04$$

3. *t-test* (Eq. (A. 2. 18)):

$$t_A = \frac{1{\cdot}77 - 1{\cdot}81}{0{\cdot}04} \sqrt{5} = -2{\cdot}24 \text{ (Method A)}$$

$$t_B = \frac{1{\cdot}77 - 1{\cdot}87}{0{\cdot}04} \sqrt{5} = -5{\cdot}59 \text{ (Method B)}$$

From the curves in Fig. A. 2. 7 the 5 per cent and 1 per cent probabilities of observing deviations equal or are larger than t for a sample with 4 degrees of freedom give:

$$t_{0{\cdot}05, 4} = 2{\cdot}78 \quad \text{and} \quad t_{0{\cdot}01, 4} = 4{\cdot}60.$$

Clearly, method A yields a result that is not significantly different from the observations as $|t_A| < t_{0 \cdot 05,4}$ and $< t_{0 \cdot 01,4}$. Method B, however, disagrees significantly with the observations as $|t_B| > t_{0 \cdot 05,4}$ and $> t_{0 \cdot 01,4}$.

Example (h)

A sample of ten measurements of length of a certain mechanical part resulted in mean $\overline{X} = 5 \cdot 3$ mm and standard deviation $S = 0 \cdot 08$ mm. Find the 95 per cent and 99 per cent confidence limits for the actual length, i.e.

$$\overline{X} \pm t_{\alpha, \nu} S / \sqrt{n}$$

Here the number of degrees of freedom $\nu = n - 1 = 9$.

95 per cent Confidence limits: $\alpha = 0 \cdot 05$;

$$t_{0 \cdot 05,9} = 2 \cdot 26,$$

$$5 \cdot 31 \pm \frac{2 \cdot 26 \times 0 \cdot 08}{\sqrt{9}} = 5 \cdot 31 \pm 0 \cdot 06$$

99 per cent Confidence limits: $\alpha = 0 \cdot 01$;

$$t_{0 \cdot 01,9} = 3 \cdot 25,$$

$$5 \cdot 31 \pm \frac{3 \cdot 25 \times 0 \cdot 08}{\sqrt{9}} = 5 \cdot 31 \pm 0 \cdot 09$$

A.2.12   The confidence limits in linear regression

Further to section A.2.10, the constants C_0 and C_1 for a least squares line according to Eq. (A.2.17) are:

$$C_0 = \frac{\Sigma Y_k \Sigma X_k{}^2 - \Sigma X_k \Sigma X_k Y_k}{n \Sigma X_k{}^2 - (\Sigma X_k)^2}$$

$$C_1 = \frac{n \Sigma X_k Y_k - \Sigma X_k \Sigma Y_k}{n \Sigma X_k{}^2 - (\Sigma X_k)^2}$$

In order to establish the confidence limits (boundaries), say at the level $100(1 - \alpha)$ per cent, for the linear regression one calculates first the standard deviation of the residuals

$$S_y{}^2 = \frac{1}{n - 2} \sum_1^n [Y_k - (C_0 + C_1 X_k)]^2, \tag{A.2.19}$$

where $n - 2$ appears because two degrees of freedom have already been used for calculating C_0 and C_1.

Now, since the total number of points used for the purpose of regression may be typical of a small sample, i.e. less than 20, it is proper to use the t-distribution rather than the normal distribution for evaluating the confidence boundaries. Their equation is expressed here in terms of the variables $\hat{Y} = f(\hat{X})$. The confidence boundaries are

$$\hat{Y} = C_0 + C_1\hat{X} \pm t_{\alpha,\,n-2}\, S_y \left[\frac{1}{n} + \frac{(\hat{X} - \overline{X})^2}{\displaystyle\sum_1^n (X_k - \overline{X})^2} \right]^{1/2}, \qquad (A.2.20)$$

Equation (A.2.20) is quoted without derivation; it defines two separate boundaries: one below and one above the linear regression $Y = C_0 + C_1 X$. Note that the confidence limits are smallest at the point where $\hat{X} = \overline{X}$. A typical formation of such confidence boundaries is illustrated in Fig. A.2.8.

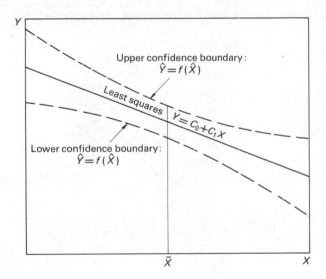

FIG A.2.8   Confidence limits for linear regression using $t_{\alpha,\,\nu}$ significance values

Similarly, the confidence boundaries can be evaluated for higher order regressions as well. However, the problem becomes then algebraically quite complex and tedious and numerical procedures based on computer facilities are best advised in those circumstances. For particulars consult more advanced textbooks (e.g. Guttman et al.[2])

A. 2. 13   Closing remarks

For obvious reasons, this appendix fails to deal with many other important aspects of error analysis and random data evaluation. No, or but casual, mention is made about such problems as: (i) Correlated or interdependent observations and errors; (ii) Weighted measurements where some observations are trusted more than the others and therefore are given higher weighting factors; (iii) goodness-of-fit tests. Neither does this appendix explore the wide scope of other useful distributions (Poisson, bionomial etc., nor does it emphasise strongly enough the role of the probability theory in processing and interpreting experimental observations. The reader is hereby encouraged to further his horizons in this area by studying the subject matter beyond the depth of the present exposition. A thorough ability of analysing and interpreting experiments and their inherent errors should be recognised as a worthwhile personal investment.

References

1. C. CHATFIELD. *Statistics for Technology,* Penguin Books, 1970.
2. I. GUTTMAN, S. S. WILKS and J. S. HUNTER. *Introductory Engineering Statistics,* Wiley, 1971.

Appendix 3   Notation, units, laws, formulae, properties and constants

Although Imperial and other units will be encountered in laboratory work, SI Units will generally be used.

The seven primary units are listed in Table A3.1, together with some special derived units. Fractions and multiples, and some conversion factors are also listed.

TABLE A3.1   SI units and Imperial units

SI units

The seven primary units

Quantity	Unit	Symbol
length	metre	m
mass	kilogram	kg
time	second	s
electric current	ampere	A
thermodynamic temperature	kelvin	K
luminous intensity	candela	cd
amount of substance	mole	mol

Some special derived units

Quantity	Unit	Symbol	Definition
plane angle	radian	rad	$180/\pi$ degrees
frequency	hertz	Hz	s^{-1}
force	newton	N	$kg\ m/s^2$
pressure, stress	pascal	Pa	N/m^2
dynamic viscosity	pascal second	Pa s	Ns/m^2

TABLE A3.1 (contd.)

SI units

Some special derived units

Quantity	Unit	Symbol	Definition
work, energy, quantity of heat	joule	J	Nm
power, heat flow rate	watt	W	J/s
electric charge	coulomb	C	As
electric potential difference	volt	V	W/A
electric resistance	ohm	Ω	V/A
electric conductance	siemens	S	A/V
electric capacitance	farad	F	As/V
magnetic flux	weber	Wb	Vs
inductance	henry	H	Vs/a
magnetic flux density	tesla	T	Wb/m^2

Fractions and multiples

Fraction	Prefix	Symbol	Multiple	Prefix	Symbol
10^{-1}	deci	d	10	deca	da
10^{-2}	centi	c	10^2	hecto	h
10^{-3}	milli	m	10^3	kilo	k
10^{-6}	micro	μ	10^6	mega	M
10^{-9}	nano	n	10^9	giga	G
10^{-12}	pico	p	10^{12}	tera	T
10^{-15}	femto	f			
10^{-18}	atto	a			

Imperial units

The six primary units are

Quantity	Unit	Symbol
length	foot	ft
mass	pound	lbm
time	second	s
electric current	ampere	A
temperature	degree Rankine	°R
luminous intensity	candela	cd

TABLE A3.1 (contd.)

Imperial units

Some special derived units

Quantity	Unit	Symbol	Definition
force	pound	lbf	$(1/g_0)$ lbm ft/s^2
work, energy	foot pound	ft lbf	ft lbf
quantity of heat	British thermal unit	Btu	J ft lbf
power	horsepower	hp	550 ft lbf/s

Notes

1. $(1/g_0)$ is the constant of proportionality in Newton's second law of motion so that

$$1 \text{ lbf} = \frac{1 \text{ lbm} \times 32 \cdot 174 \text{ ft/s}^2}{g_0}$$

Therefore $g_0 = 32 \cdot 174 \dfrac{\text{lbm ft}}{\text{lbf s}^2}$

2. J is Joule's equivalent and relates heat units to work units in the imperial system so that

   1 Btu = 778 ft lbf

   or J = 778 ft lbf/Btu

Conversion factors (slide rule accuracy)

	Imperial Unit		SI Unit
length	1 in	=	$2 \cdot 540$ cm
	1 ft	=	$0 \cdot 3048$ m
	1 mile	=	$1 \cdot 609$ km
area	1 in^2	=	$6 \cdot 452$ cm^2
	1 ft^2	=	$0 \cdot 0929$ m^2
volume	1 in^3	=	$16 \cdot 39$ cm^3
	1 ft^3	=	$0 \cdot 02832$ m^3
velocity	1 ft/s	=	$0 \cdot 3048$ m/s
	1 mile/h	=	$1 \cdot 609$ km/h
mass	1 lbm	=	$0 \cdot 4536$ kg
	1 slug	=	$14 \cdot 59$ kg
density	1 lbm/ft^3	=	$16 \cdot 02$ kg/m^3
	1 slug/ft^3	=	$515 \cdot 4$ kg/m^3

TABLE A3.1 (contd.)

Conversion factors (slide rule accuracy)

	Imperial Unit		SI Unit
force	1 lbf	=	4·448 N
pressure	1 lbf/in^2	=	6·895 kPa
	1 in H$_2$O	=	249·1 Pa
	1 in Hg	=	3·386 kPa
energy	1 ft lbf	=	1·356 J
	1 Btu	=	1·055 kJ
power	1 hp	=	745·7 W

Although notation will vary from person to person, the notation and associated units given in Table A3.2 are widely used.

Several fundamental laws and derived relations, which will prove useful during experimental work, are summarised on subsequent pages. Finally, some standard values, properties and constants are quoted.

TABLE A3.2   List of commonly used symbols

Symbol	Quantity	USUAL UNITS SI	Imperial
A	area	m^2	ft^2
a	sonic velocity	m/s	ft/s
B	intensity of magnetic field	T	
b	breadth	m	ft
C	capacitance	F	
C_p	specific heat at constant pressure	kJ/kgK	Btu/lbm°R
C_v	specific heat at constant volume	kJ/kgK	Btu/lbm°R
d	diameter, depth	m	ft
	distance between plates of condenser	m	
E	energy	kJ	ft lbf, Btu
	modulus of elasticity	GPa	lbf/in^2
	intensity of electric field	V/m	

TABLE A3.2 (contd.)

Symbol	Quantity	USUAL UNITS SI	Imperial
e	specific energy	kJ/kg	ft lbf/lbm
F	force	N	lbf
f	acceleration	m/s^2	ft/s^2
	frequency	Hz	s^{-1}
G	shear modulus	GPa	lbf/in^2
g	gravitational acceleration	m/s^2	ft/s^2
H	enthalpy,	kJ	Btu
	angular momentum	kg m^2/s	lbm ft^2/s
	height	m	ft
h	specific enthalpy	kJ/kg	Btu/lbm
	heat transfer coefficient	W/m^{2}K	Btu/ft^{2}h°R
I	second moment of area	m^4	ft^4
	moment of inertia	kg m^2	lbm ft^2
	electric current	A	
K	dielectric constant	—	
k	radius of gyration	m	ft
	thermal conductivity	W/mK	Btu/ft h°R
L	self-inductance	H	
l	length	m	ft
M	Mach number	—	—
	moment	Nm	lbf ft
	mutual inductance	H	
$\overline{M}$	molecular mass	—	—
m	mass	kg	lbm
N	angular velocity	rev/min	rev/min
n	frequency	Hz	s^{-1}
P	stagnation pressure	kPa	lbf/in^2
p	pressure	kPa	lbf/in^2
Q	heat transfer	kJ	Btu
	volume flow rate	m^3/s	ft^3/s
	isolated electric point charge	C	
q	heat transfer per unit mass	kJ/kg	Btu/lbm
	electric point charge	C	
R	gas constant	kJ/kgK	Btu/lbm° R
	resistance	Ω	
$\overline{R}$	molal gas constant	kJ/kg-molK	Btu/lb-mol°R

TABLE A3.2 (contd.)

Symbol	Quantity	USUAL UNITS SI	Imperial
r	radius	m	ft
	distance between electric charges	m	
S	entropy	kJ/K	Btu/°R
	surface area	m^2	ft^2
s	specific entropy	kJ/kgK	Btu/lbm°R
	distance	m	ft
T	torque	Nm	lbf ft
	temperature,	K, °C	°R, °F
	kinetic energy	J	ft lbf
t	time	s	s
U	internal energy strain energy	kJ	ft lbf, Btu
u	specific internal energy	kJ/kg	Btu/lbm
V	velocity	m/s	ft/s
	total volume	m^3	ft^3
	shear force	N	lbf
	potential difference	V	
v	specific volume	m^3/kg	ft^3/lbm
W	work, work transfer	J	ft lbf, Btu
W_S	shaft work	J	ft lbf, Btu
x, y, z	Cartesian coordinates	—	—
y	depth	m	ft
z	expansion	m	ft
α	angular acceleration	rad/s^2	rad/s^2
	angle	rad	rad
	coefficient of linear expansion	m/mK	in/in°R
γ	ratio of specific heats C_p/C_v	—	—
	shear strain	—	—
δ	deflection	m	ft
	elongation	m	ft
ϵ	linear strain	—	—
	electromotive force	V	
ϵ_0	permittivity of free space	pF/m	
η	efficiency	—	—
θ	angle	rad	rad

TABLE A3.2 (contd.)

Symbol	Quantity	USUAL UNITS SI	Imperial
μ	dynamic viscosity	Pa s	slug/ft s
μ_0	permeability of free space	Wb/Am	
ν	kinematic viscosity	m^2/s	ft^2/s
	Poisson's ratio	—	—
ρ	density	kg/m^3	$slug/ft^3$
	charge density	C/m^3	
	resistivity	Ωm	
σ	direct stress	Pa	lbf/in^2
	surface charge density	C/m^2	
τ	shear stress	Pa	lbf/in^2
ϕ	flux of magnetic field	Wb	
ω	angular velocity	rad/s	rad/s
	circular frequency	rad/s	rad/s

Superscripts

·   first derivative with respect to time $\dfrac{d}{dt}$

··   second derivative with respect to time $\dfrac{d^2}{dt^2}$

^   unit vector
    molal quantity

Subscripts

_   vector quantity

      With Imperial units 'inch' and 'foot' can obviously be interchanged (using the appropriate factor of 12) and 'pound mass' (lbm) and 'slug' ($=32\cdot174$ lbm) can also be interchanged.

Fundamental laws

Second law of motion

Force = rate of change of linear momentum with time

$$F = \frac{d}{dt}(mV) \tag{A3.1}$$

Torque = rate of change of angular momentum with time

$$T = \frac{dH}{dt} \tag{A3.2}$$

Conservation of mass

Providing relativistic effects are negligible, the change in mass of an open system is equal to the net mass of fluid crossing the boundary of the system.   (A3.3)

First law of thermodynamics

For a closed system, which is taken through any thermodynamic cycle, the cyclic integral of the heat transfers is equal to the cyclic integral of the work transfers.

$$\oint dW = \oint dQ \tag{A3.4}$$

Second law of thermodynamics

(a) *Kelvin–Planck statement.* It is impossible to construct a device which will operate in a cycle and produce no effect other than the raising of a weight (work transfer) and the exchange of heat with a single reservoir.   (A3.5(a))

(b) *Clausius statement.* It is impossible to construct a device which operates in a cycle and produce no effect other than the transfer of heat from a cooler body to a hotter body.   (A3.5(b))

Equation of state for a perfect gas

For some pure substances, known as perfect or ideal gases, the product or pressur and specific volume is proportional to the absolute temperature

$$pv = RT \tag{A3.6}$$

Hooke's law for elastic materials

For an elastic material direct stress is proportional to direct strain.

$$\text{In one dimension } \sigma = E\epsilon \tag{A3.7}$$

For an elastic material shear stress is proportional to shear strain.

$$\text{In one dimension } \tau = G\gamma \tag{A3.8}$$

Coulomb's law of electrostatic forces

The electrostatic force between two point charges is directly proportional to the product of their magnitudes and inversely proportional to the square of their separation.

$$F = \frac{q_1 q_2}{4\pi \epsilon_0 r^2} \tag{A3.9}$$

Ohm's law of resistance

For electrically conducting materials, the rate of flow of electric charge (or current) is directly proportional to the electric field which, in turn, is directly proportional to the applied potential difference

$$\frac{dq}{dt} = I \; \alpha \; V \text{ or } V = IR \qquad \text{(A3.10)}$$

(Although this was an experimentally derived law it can in fact be deduced from the directed motion of electrons under the influence of an electric field).

Biot–Savart law for magnetic fields

For a current element of length δl which carries a current I the magnetic field intensity due to the element is given by

$$\delta B = \frac{\mu_0 I \delta l \, \sin \theta}{4\pi r^2} \qquad \text{(A3.11)}$$

where θ is the angle between the radius vector r and the current element.

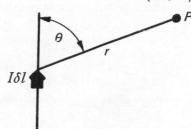

Faraday's law for induced electromotive force

The induced e.m.f. is proportional to the time rate of change of the flux of the magnetic field surrounded by the circuit

$$\epsilon = - \frac{d\phi}{dt} \qquad \text{(A3.12)}$$

Definitions

Poisson's ratio for elastic materials

For an elastic material the ratio of lateral strain to longitudinal strain is a constant

$$\nu = - \frac{\text{lateral strain}}{\text{longitudinal strain}} \qquad \text{(A3.13)}$$

Relation between elastic constants

$$G = \frac{E}{2(1 + \nu)} \tag{A3.14}$$

Work

The work done by a force is the product of that force and the displacement of the point of action of that force in the direction of the force.

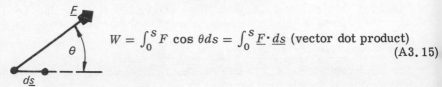

$$W = \int_0^S F \cos \theta \, ds = \int_0^S \underline{F} \cdot \underline{ds} \text{ (vector dot product)} \tag{A3.15}$$

Electric field

The vector force acting on a unit positive charge at any point in space defines the electric field at that point

$$\text{Thus } \underline{E} = \frac{F}{q} \tag{A3.16}$$

Electrostatic potential

The electrostatic potential (ϕ_p) at a point is the work done per unit charge against the field in bringing a small positive charge from infinity to that point

$$\text{Thus } \phi_p = \frac{W}{q} = - \int_\infty^p \underline{E} \cdot d\underline{l} \tag{A3.17}$$

Electron volt

An electron volt is the energy gained by a particle of one electronic charge which is accelerated through one volt potential difference

$$\text{Thus } 1 \text{ e.v.} = -1 \cdot 602 \times 10^{-19} \text{ joules}$$

Capacitance

The capacitance measures the amount of charge necessary to increase the potential of a conductor by one volt

$$\text{Thus } C = \frac{Q}{V} \tag{A3.18}$$

(Generally μF or pF are most appropriate fractional units)

Dielectric constant

The dielectric constant is defined as the ratio

$$K = \frac{\text{capacitance of a capacitor with dielectric between the plates}}{\text{capacitance of the same capacitor without dielectric}}$$

Electromotive force (e.m.f.)

The electromotive force ϵ is the work per unit charge done by a nonelectrostatic source of energy

$$\epsilon = \frac{dW}{dq} \tag{A3.19}$$

Mutual inductance

Mutual inductance (M) is the constant of proportionality which relates the e.m.f. induced in one circuit by the rate of change of current in a second circuit

$$\epsilon_a = - M_{ab} \frac{dI_b}{dt} \text{ and } M_{ab} = M_{ba} = M \tag{A3.20}$$

Self inductance

Self inductance (L) is the constant of proportionality which relates the e.m.f. induced in a single circuit by the rate of change of current in the same circuit.

$$\epsilon = - L \frac{dI}{dt} \tag{A3.21}$$

Derived relations

Velocity and Acceleration along a straight path

$$f = \dot{V} = \ddot{s} = \frac{dV}{dt} = \frac{ds}{dt}\frac{dV}{ds} = V\frac{dV}{ds} = \frac{d^2s}{dt^2} \tag{A3.22}$$

hence $V_2 = V_1 + ft;$ $\qquad V_2{}^2 = V_1{}^2 + 2fs;$ $\qquad \frac{ds}{dt} = V_1 + ft$

$$s = V_1 t + \tfrac{1}{2} ft^2$$

where V_1 is the initial velocity in each case and s is the distance travelled (A3.23)

Velocity and acceleration along a curved path

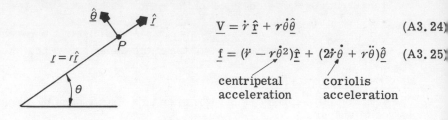

$$\underline{V} = \dot{r}\,\hat{\underline{r}} + r\dot{\theta}\hat{\underline{\theta}} \qquad (A3.24)$$

$$\underline{f} = (\ddot{r} - r\dot{\theta}^2)\hat{\underline{r}} + (2\dot{r}\dot{\theta} + r\ddot{\theta})\hat{\underline{\theta}} \qquad (A3.25)$$

centripetal   coriolis
acceleration   acceleration

Motion of a constant mass

Equation A3.1 becomes $F = m\dfrac{dV}{dt} = mf = m\ddot{s}$ (A3.26)

Rotation of a rigid body

Equation A3.2 becomes $T = I\dfrac{d\omega}{dt} = Ia = I\ddot{\theta}$ (A3.27)

Energy

Kinetic. For a constant mass A3.26 gives $F = m\ddot{s}$ and for $\theta = 0$ A3.15 becomes

$$dW = Fds = m\ddot{s}\,ds = m\frac{d\dot{s}}{dt}\,ds = m\,d\dot{s}\,\frac{ds}{dt}$$

$$dW = m\dot{s}\,d\dot{s} = d(\tfrac{1}{2}m\dot{s}^2) \qquad (A3.28)$$

work done = change of kinetic energy.

Potential. In a gravitational field $F = mg$

If z measured in reverse direction to gravitational attraction then

$$W = \int \underline{F}\cdot d\underline{s} = \int_0^z mg\,dz = mgz \qquad (A3.29)$$

work done = change of potential energy.

Power

Power is the rate of doing work $= \dot{W}$

$$= \frac{d}{dt}\int \underline{F}\cdot d\underline{s}$$

If F is constant and angle θ is zero then

$$\dot{W} = F\frac{ds}{dt} = FV$$

For an engine brake $V = \omega r$ and $F = \dfrac{T}{r}$

$$\dot{W} = T\omega = \frac{2\pi NT}{60}$$

or $\qquad \dot{W} = \dfrac{2\pi NFr}{60} = \dfrac{FN}{k}$ $\qquad\qquad$ (A3. 30)

where k is the brake constant $= \dfrac{60}{2\pi r}$

N is the speed in rev/min
T is the torque in Nm
$\dot{W}$ is the power in W (or J/s)
F is the load in N

Mohr stress circle

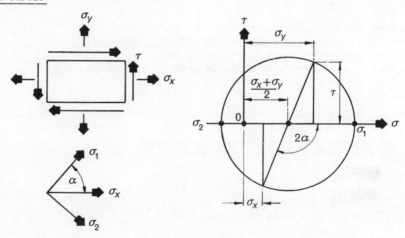

$$\sigma_{1,\,2} = \left(\frac{\sigma_x + \sigma_y}{2}\right) \pm \sqrt{\left(\frac{\sigma_y - \sigma_x}{2}\right)^2 + \tau^2}$$

$$\tan 2\alpha = -\frac{2\tau}{\sigma_y - \sigma_x} \qquad\qquad \text{(A3. 31)}$$

Elastic bending of beams

$$\frac{M}{I} = \frac{\sigma}{y} = \frac{E}{R} \qquad\qquad \text{(A3. 32)}$$

where σ = bending stress at distance y from neutral axis
$\quad\quad M$ = bending moment at section considered
$\quad\quad R$ = change in radius of curvature of beam

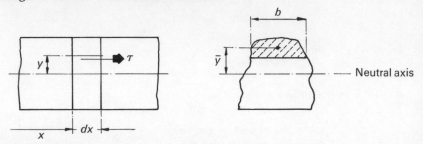

Shear stress distribution at distance x along the beam

$$\tau = \frac{VA\bar{y}}{Ib}$$

(A3.33)

where τ = shear stress at distance y from neutral axis
$\quad\quad V$ = shear force at distance x along the beam
$\quad\quad A$ = shaded area
$\quad\quad \bar{y}$ = distance of centroid of area A from the neutral axis.

Torsion formulae

Round shafts

$$\frac{T}{J} = \frac{G\theta}{l} = \frac{\tau}{r}$$

(A3.34)

where T = torque
$\quad\quad J$ = polar second moment of area
$\quad\quad \theta$ = angle of twist
$\quad\quad \tau$ = shear stress at radius r

Thin walled tube

$$\tau = \frac{T}{2At}$$

(A3.35)

and $\quad\quad\quad \dfrac{\theta}{l} = \dfrac{T}{4A^2G} \oint \dfrac{ds}{t}$

(A3.36)

where A = enclosed area
$\quad\quad t$ = wall thickness
$\quad\quad ds$ = short length of perimeter

TABLE A3. 3   SECTION PROPERTIES

Second Moment of Area

(a) Rectangle about XX $\dfrac{bd^3}{3}$

(b) Rectangle about CC $\dfrac{bd^3}{12}$

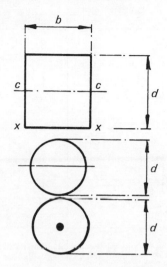

(c) Circle about diameter $\dfrac{\pi d^4}{64}$

(d) Polar moment of circle $\dfrac{\pi d^4}{32}$

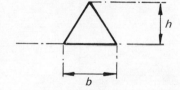

(e) Triangle about base $\dfrac{bh^3}{12}$

(f) Triangle about axis through centre of area $\dfrac{bh^3}{36}$

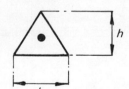

Parallel axis theorem $I_{XX} = I_G + Ah^2$

where XX is axis, distance h from centre G of area A

Perpendicular axis theorem $I_Z = I_X + I_Y$

TABLE A3.4   SHEAR FORCE AND BENDING MOMENT DIAGRAMS

Beam	Shear force distribution	Bending moment distribution	Maximum deflection
W ; l	W	0 ; 0 ; Wl	$\dfrac{1}{3}\dfrac{Wl^3}{EI}$
w per unit length ; l	wl ; 0 ; 0	0 ; 0 ; $\dfrac{wl^2}{2}$	$\dfrac{1}{8}\dfrac{wl^4}{EI}$
W (mid-span) ; l	$\dfrac{W}{2}$; 0 ; $\dfrac{W}{2}$	0 ; $\dfrac{Wl}{4}$; 0	$\dfrac{1}{48}\dfrac{Wl^3}{EI}$
w per unit length ; l	$\dfrac{wl}{2}$; 0 ; 0 ; $\dfrac{wl}{2}$	0 ; $\dfrac{wl^2}{8}$; 0	$\dfrac{5}{384}\dfrac{wl^4}{EI}$
W (mid-span) ; l	$\dfrac{W}{2}$; 0 ; $\dfrac{W}{2}$	0 ; $\dfrac{Wl}{8}$; 0 ; $\dfrac{Wl}{8}$	$\dfrac{1}{192}\dfrac{Wl^3}{EI}$
w per unit length ; l	$\dfrac{wl}{2}$; 0 ; $\dfrac{wl}{2}$	$\dfrac{wl^2}{24}$; 0 ; 0 ; $\dfrac{wl^2}{12}$	$\dfrac{1}{384}\dfrac{wl^4}{EI}$

Bernoulli's equation

For the steady flow of an inviscid incompressible fluid along a streamline, equation A3.1 gives

$$\frac{p}{\rho} + \frac{V^2}{2} + gz = \text{constant} \tag{A3.37}$$

Continuity

For steady flow, the rate of fluid mass flowing across any section of a stream tube remains constant (derived from A3.3)

$$\dot{m} = \rho A V = \text{constant} \tag{A3.38}$$

Hydrostatic pressure

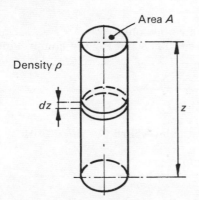

Area A

Density ρ

dz

z

From A3.1 Force $F =$ weight of fluid column = mass $\times$ acceleration =

$$\int_0^z (\rho A\,dz)\, g = \rho A g z$$

Pressure due to column =

$$\frac{F}{A} = \rho g z \tag{A3.39}$$

Centre of pressure

$$y_p = \frac{{}_A\!\int y^2 dA}{{}_A\!\int y\,dA} = \frac{\text{second moment of area}}{\text{first moment of area about surface}} \tag{A3.40}$$

Height of metacentre from centre of buoyancy

$$MB = \frac{I}{V} = \frac{\text{second moment of water line area}}{\text{total displaced volume}} \tag{A3.41}$$

Flowmeters—orifices, nozzles and venturi meters

For incompressible flow A3.37 and A3.38 with modification for real viscous fluids give equations of the form

$$\dot{m} = k\sqrt{h} \tag{A3.42}$$

where k is a constant related to the dimensions of the flowmeter, viscous effects, and the densities of both the moving fluid and the manometer fluid, and h is the pressure difference (or 'head') across the flowmeter often expressed as the length of a column of manometric fluid (see equation A3.39)

For compressible flow of perfect gases equations A3.6, A3.38 and A3.49 give equations of the form

$$\dot{m} = k\epsilon\sqrt{\frac{hp}{T}} \tag{A3.43}$$

where k is a constant for a particular flowmeter and manometer
   h is the differential head
   p is the upstream pressure
   T is the upstream temperature (absolute)
   ϵ is an expansion factor used to account for the compressibility.

Pitot tubes for flow measurement

Equation A3.37 gives $p_0 - p = \frac{1}{2}\rho V^2$ $\tag{A3.44}$

where p_0 is the stagnation pressure and
   p is the static pressure at a point where the velocity is V

$$\text{Hence } V = \sqrt{\frac{2(p_0 - p)}{\rho}} \tag{A3.45}$$

$$\text{and } \quad \dot{m} = \int_{\text{area}} \rho V dA \tag{A3.46}$$

where p_0 is the stagnation pressure in Pa
   p is the static pressure in Pa
   ρ is the density in kg/m^3
   V is the velocity in m/s

Sonic velocity

Equations A3.1 and A3.38 give

$$a = \sqrt{\left(\frac{dp}{d\rho}\right)_s} \tag{A3.47}$$

for liquids and gases.

With equation A3.6 for perfect gases (only)

$$a = \sqrt{\gamma R T} \qquad (A3.48)$$

where a is sonic velocity in m/s
    R is the gas constant in J/kgK
    T is the absolute temperature in K

Steady flow energy equation

Applying equation A3.4 to an open system with steady flow gives

$$\dot{Q} - \dot{W}_S = \dot{m} \Delta \left(h + \frac{V^2}{2} + gz \right) \qquad (A3.49)$$

Entropy

Entropy is a property of a pure substance which results from A3.5 such that the change in entropy

$$ds = \frac{dQ \text{ reversible}}{T} \qquad (A3.50)$$

Thermodynamic work

For a reversible (frictionless, quasistatic) process equation A3.15 gives displacement work

$$W = \int p \, dV \qquad (A3.51)$$

where dV is the volume change.

Electric field

From A3.9 add A3.16

$$\underline{E} = \frac{1}{4\pi\epsilon_0} \sum_i \frac{q_i}{r_i^2} \hat{r}_i \qquad (A3.52)$$

$$\int_{c.s} \int \underline{E} \cdot d\underline{S} = \sum_{c.v}^{i} \frac{q_i}{\epsilon_0} = \iiint_{c.v} \frac{\rho}{\epsilon_0} \, dV$$

(Gauss' Flux Theorem)   (A3.53)

From A3.17

$$\oint \underline{E} \cdot d\underline{l} = 0 \qquad (A3.54)$$

Potential difference

From A3.17

$$V = - \int_A^B \underline{E} \cdot d\underline{l} \tag{A3.55}$$

For a conducting sphere A3.52 gives

$$V = \frac{Q}{4\pi\epsilon_0 r} \tag{A3.56}$$

Capacitance

From A3.18 and A3.56 for a sphere

$$C = 4\pi\epsilon_0 r \tag{A3.57}$$

For a parallel plate condenser A3.53 gives

$$E = \frac{\sigma}{\epsilon_0} = \frac{Q}{\epsilon_0 A} \tag{A3.58}$$

Using A3.55 and A3.58 gives

$$V = \frac{Qd}{\epsilon_0 A} \tag{A3.59}$$

Thus $C = \dfrac{A\epsilon_0}{d}$ (A3.60)

For capacitors in series $\dfrac{1}{C} = \sum_i \dfrac{1}{C_i}$ (A3.61)

For capacitors in parallel $C = \sum_i C_i$ (A3.62)

Resistance

From A3.10 $R = \rho \dfrac{l}{A}$ (A3.63)

where ρ = resistivity in ohm metres (Ωm)

For resistors in series $R = \sum_i R_i$ (A3.64)

For resistors in parallel $\dfrac{1}{R} = \sum_i \dfrac{1}{R_i}$ (A3.65)

Kirchhoff's rules

1. From charge conservation, the algebraic sum of the currents at a junction is zero

$$\Sigma I = 0 \qquad\qquad\qquad \text{(A3.66)}$$

2. From Ohm's Law (A3.10), the sum of the e.m.f.'s in a circuit is equal to the sum of the voltage drops in the resistance around the circuit

$$\Sigma \epsilon = \Sigma IR \qquad\qquad\qquad \text{(A3.67)}$$

Power

From A3.10 and A3.19,

$$\text{Power } \dot{W} = \frac{dW}{dt} = \frac{dW}{dq}\frac{dq}{dt} = \epsilon I \qquad\qquad \text{(A3.68)}$$

or $\quad \dot{W} = VI$ if the internal resistance of the source of e.m.f. is negligible. $\qquad\qquad$ (A3.69)

or using A3.10, $\dot{W} = I^2 R = \dfrac{V^2}{R}$ $\qquad\qquad$ (A3.70)

Magnetic field

From A3.11 $\displaystyle\int\int_{c \cdot s} \underline{B} \cdot d\underline{S} = 0$

(Gauss' Theorem for a magnetic field) $\qquad\qquad$ (A3.71)

Magnetostatics

$$\oint \underline{B} \cdot d\underline{l} = \mu_0 \Sigma I \qquad\qquad\qquad \text{(A3.72)}$$

Inductance

For self-inductances in series $L = \displaystyle\sum_i L_i$ $\qquad\qquad$ (A3.73)

For self-inductances in parallel $\dfrac{1}{L} = \displaystyle\sum_i \dfrac{1}{L_i}$ $\qquad\qquad$ (A3.74)

TABLE A3.5   SOME STANDARD VALUES (*To accuracy required for slide-rule calculations*)

Standard Gravitational Acceleration $= 9\cdot81$ m/s^2

International Standard Atmosphere (I.S.A.)
         Pressure $= 1\cdot013$ bar $= 101\cdot3$ kN/m^2 $= 101\cdot3$ kPa
         Temperature $= 15$ °C $= 288$ K

Universal Gas Constant $\overline{R} = 8\cdot314$ kJ/kg mol K

Molal Volume $\overline{V} = 22\cdot41$ m^3/kg mol at I.S.A. pressure and 0 °C

Composition of air

	$\overline{M}$	Vol. analysis	Grav. analysis
Nitrogen (N$_2$)	28	0·79	0·767
Oxygen (O$_2$)	32	0·21	0·233

Properties of air
         Molecular mass $\overline{M} = 29$
         Specific Gas Constant $R = 0\cdot287$ kJ/kg K
         Specific Heat at Constant Pressure $C_p = 1\cdot005$ kJ/kg K
         Specific Heat at Constant Volume $C_p = 0\cdot718$ kJ/kg K
         Ratio of specific heats $\gamma = C_p/C_v = 1\cdot4$
         Thermal Conductivity $k = 0\cdot0253$ W/mK at I.S.A. conditions
         Dynamic Viscosity $\mu = 17\cdot9$ μPa s
         Density at I.S.A. conditions $\rho = 1\cdot225$ kg/m^3
         Sonic velocity at I.S.A. conditions a $= 340$ m/s

Properties of water
         Molecular mass $\overline{M} = 18$
         Specific heat at constant pressure at 15 °C $C_p = 4\cdot186$ kJ/kg K
         Thermal conductivity $k = 595$ kW/m K at 15 °C
         Dynamic viscosity $\mu = 1\cdot14$ mPa s at 15 °C
         Density at 4°C $\rho = 1000$ kg/m^3

TABLE A3.6   Some properties of some metals and alloys (*Average Values at 15 °C*)

Property	Units	Aluminium	Brass	Cast Iron	Copper	Lead	Mild Steel	Nimonic	Silver
Density	kg/m^3	2 710	8 370	7 600	8 960	11 300	7 800	8 200	10 500
Elastic modulus	GPa	70	100	115	112	16	206	200	83
Poisson's ratio	—	0·34	0·37	0·26	0·36	0·44	0·28	—	0·37
Shear modulus	GPa	26	37	46	41	6	80	88	30
Ultimate strength	MPa	120	430	210(T) 730(C)	230	15	490	1 200	290
Coefficient of thermal expansion	μm/m K	24	19	9	16	29	12	12	19
Thermal conductivity	W/m K	234	114	48	384	35	47	113	418
Melting point	K	933	1 200	1 500	1 356	600	1 600	1 650	1 234
Electrical resistivity	nΩm	28	62	600	17	207	160	1 200	16

T = Tensile   C = Compressive

These properties depend on impurities, alloying elements, heat treatment, method of production, temperature, environment, etc. If required, more detailed and accurate information can be obtained from British Standard Specifications, Manufacturers Specifications, or textbooks such as *Physical and Chemical Constants* by G. W. C. Kaye and T. H. Laby (Longman).

TABLE A3.7 Some properties of some non-metals (*Average Values*)

Property	Units	Asbestos	Carbon	Concrete	Glass	Plastics Thermoplastic (Polyethylene)	Plastics Thermoset (Epoxy)	Wood Oak	Wood White pine
Density	kg/m^3	2 400	1 500	2 000	2 500	930	1 115	700	500
Elastic modulus	GPa		5		70	0·3	2·4	10·3	10·3
Poisson's ratio	—			0·10	0·25				
Coefficient of thermal expansion	μm/m K		2·4	12	8	150	60		
Thermal conductivity	W/m K	0·2	1·6	1·0	1·0	0·38	0·19	0·17	0·12
Electrical resistivity	TΩm $\mu\Omega$m		14		>0·01	>100	1·0		

Index

Abbreviations, 121
Accuracy, 34, 48, 74, 75, 91
Arithmetic mean, 131
Average, 83

Bending of beams, formulae for, 165
Bernoulli equation, 56, 167
Bourdon gauge, 54
Buckingham Theorem, 129

Calibration, 36
Capacitance, 171
Checking data, 17
Clock, 40
Communication, 4, 116, 125
Confidence,
    level, 149
    limits, 150
Conservation of mass, 160
Conversion factors, 155
Curve fitting, 3, 12, 97, 143

Data, recording, 17
Design, 9
    of experiment, 21
Derived relationships, 163-4
Deviation,
    probable, 134
    standard, 132

Dial gauge, 41
Dimensions, 128
Dimensional analysis, 2, 21
Discrimination, 34
Displacement, measurement, 41, 68
Dynamometer, 49

Entropy, 120
Equipment,
    compendium of, 37-72
    selection, 11, 33
Errors, 3, 36, 74-7, 85, 86, 142
Experiment,
    illustrative, 7, 8
    in design, 7, 9
    investigative, 7, 8
    superfluous, 22
Experimental method, 6
Extensometer, 41
Extrapolation, 3, 19, 86, 110, 111,
    113-14

Flatness factor, 137
Flowmeter, 61-2, 70, 168
Force measurement, 49, 69
Frequency,
    counter, 40
    of occurrence, 131
Fundamental Laws, 159-63

Gaussian distribution, 133-7
Goodness of fit, 138
Graphs, 23, 81, 123
    log-log, 17, 103

Histogram, 131
Hot wire probe, 58, 59
Hydraulic gauge, 49, 55

Independent variables, 24, 25
Integration, 106
    graphical, 108
    numerical, 108
Interference from instruments, 35, 86
Interpolation, 3, 110, 111, 119
Imperial units, 154
Instruments, 2, 11, 33

Kinetic energy, 164

Laser, 59, 60
Least squares, 144, 145
Length measurement, 37, 68
Linear variable differential trans-
    former, 45, 49, 55
Load cell, 49

Mach number, 56
Mass,
    balance, 48
    measurement, 48, 69
Maxima, minima, 16, 25, 26, 97
Measurement,
    multisample, single sample, 78
    technique, 9
Metacentre, 168
Micromanometer, 33, 34, 52, 53
Micrometer, 38, 39
Mistakes, 77, 80, 87, 88, 90, 93
Modulus of elasticity, 13, 175
Most probable value, 131, 146

National Physical Laboratory, 36, 58,
    59

Ohm's Law, 161
Optical pyrometer, 67
Oscilloscope, 40
Orifice, 61, 168
Origin, 94
    false, 87, 88

Piezoelectric devices, 47, 48, 49
Plotting,
    as-you-go, 16, 81
    graphs, 16, 81
Poisson's ratio, 161
Population, 132
Potential energy, 164
Power, 164
Precision, 34, 75, 134
Preliminary work, 81
Pressure,
    measurement, 34, 52-60, 70, 169
    transducer, 34, 55
Probability, 134
Procedure, 11, 12
Properties,
    air, 174
    metals, 175
    non-metal solids, 176
    water, 174

Range, 11, 34
Recorder, 40
Recording data, 17, 40
Rejection criteria, 134, 136
Reports, 3, 116, 117
Resistance,
    electrical, 161, 171
    strain gauge, 45, 49, 55
    Thermometer, 64
Root mean square, 132
Rounding off, 75
Rule, 8, 37

Scale,
    instrument, 81, 82
    model, 29, 31
Scatter, 15, 17, 18, 98, 99
Section properties, 167
Sensitivity, 34
Significance, 134, 136, 147, 148
SI units, 153

Skewness, 140
Slope, 96
Speaking, 126
Spring balance, 48
Standards, 36
Strain,
    gauge, 35, 45, 49, 50, 55
    measurement, 68
Stroboscope, 41
Surface tension effects, 52-3
Symbols, 118

t-distribution, 148
Tables, 85, 88
Tachometer, 41
Temperature measurement, 62, 71
Theodolite, 41, 43
Thermocouples, 33, 65-7
Thermometers, 62-7

Time, measurement of, 40, 68
Torque, measurement of, 48, 51, 69
Torsion, 166
True Value, 74
Transducers, 40-9

Units, 119, 125, 153-9
    Imperial, 154
    SI, 153

Variables, choice of, 10, 28
Variance, 132
Vernier, 37

Wheatstone bridge, 34, 45, 46, 58, 59,
    64, 65